FOREST POLICY AND TRIBAL DEVELOPMENT

FOREST POLICY AND TRIBAL DEVELOPMENT

A Study of Maharashtra

RUCHA S. GHATE
Department of Economics
Nagpur University, Nagpur

Foreword by
SHARAD KULDARNI
Director
Centre for Tribal Conscientisation, Pune

CONCEPT PUBLISHING COMPANY, NEW DELHI 110059

The Indian Council of Social Science Research gave the financial support towards the publication of this work. However, the responsibility for the contents of the book — facts, opinions, conclusions, etc. — is entirely that of the author and not the ICSSR.

ICSSR Consultant : Prof. B.K. Roy Burman

ISBN 81-7022-418-7

First Published 1992

Published and Printed by
Ashok Kumar Mittal
Concept Publishing Company
A/15-16, Commercial Block, Mohan Garden
NEW DELHI 110059 (India)

Lasertypeset by
Microtech Advance Printing Systems (Pvt.) Ltd.
H-13, Bali Nagar
NEW DELHI 110015

FOREWORD

In general the changing relationship between forest policy and tribal communities has received scant attention at the hands of the economists. Dr. Rucha Ghate deserves congratulations for entering this somewhat virgin land. She has attempted to study the changing economic conditions of tribal communities as a consequence of forest policies adopted over a period of time with special reference to certain areas in Maharashtra.

Circulation of the Draft Forest Bill in 1980 generated discussion on the evolution of forest policy. It was pointed out that the introduction of a scientific forest policy for the first time after the advent of the British rule was a myth propagated by the foresters trained under the British rule. It is now admitted that before the advent of the British rule in India, the forests were mainly under the control of the community and there were customary rules regulating the use of different types of forests. Dr. Ghate has mentioned various social and cultural institutions developed by the people in various forest areas which enabled prudent utilisation of forest produce.

The advent of the British rule marks an important watershed in the regulation of forests in India. Foresters under the guidance of Dietrich Brandis did collect scientific data on different species of trees in India and the process was followed by distinguished botanists from the Forest Department. Secondly, it also led to the rise of systematic exploitation of forest resources by the government for the purposes of collection of revenue. Dalhousie's Memorandum on Forest Conservation issued in 1855 emphasises the importance of teak timber as a source of revenue. The emphasis on revenue from forests continued for a long time. This was rightly criticised by J.A. Voeleker, Consulting Chemist to the Royal Agricultural Society of England, who remarked that the success of the Forest Department was gauged mainly by fiscal considerations and that the Department was called upon to show a large revenue. Successive legal attempts to impose more and more restrictions on the uses of forest produce have been ably described by Dr. Ghate.

The sad part of the story is that the commercially-oriented forest policy not only continued after Independence but also gained momentum. The Forest Policy Resolution of 1952 did pay some lipservice to the cause of the amelioration of physical and climatic conditions promoting the general well-being of the people but made no reference to the dependence of the millions of members of the tribal communities on forests for their very survival. On the contrary it emphasised the need for the realisation of the maximum annual revenue in perpetuity.

The Scheduled Areas and the Scheduled Tribes Commission criticised this commercial orientation of forest policy at the cost of subsistence of tribal communities living in and near forests. It rightly pointed out that what were 'rights and privileges' of forest dwellers in 1894, became 'rights and concessions' in 1952 and only 'concessions' thereafter. However, all its recommendations found no echo in policies followed after the publication of the report.

The National Commission on Agriculture blamed the forest dwellers for growing deforestation and recommended abrogation of concessions granted to them. It came to a conclusion that free supply of forest produce to the rural population and their rights and privileges have brought destruction to the forests and so it was necessary to reverse the process. The Commission recommended clearfelling of forests and their replacement by plantations for attaining higher production per unit area, to intensify efforts to increase man made forests.

The Draft Forest Bill was based on the recommendations of the National Commission on Agriculture. However, the publication of the Bill led to vehement protests from activists working among the tribal communities, environmentalists, scientists and intellectuals. As a result, the Bill was not introduced in the Parliament.

The National Forest Policy statement adopted by the Parliament in 1988 has finally tilted the balance in favour of environmental protection and improvement. It is clearly stated that the principal aim of forest policy must be to ensure environmental stability and ecological balance including atmospheric equilibrium which are vital to sustenance of all life forms — human, animal and plant. The derivation of direct economic benefit must be subordinated to this principal aim.

This policy must be translated into practice by replacing the

Indian Forest Act of 1927, without which the new forest policy will just remain a pious statement.

Dr. Ghate has also done some field investigation. One who knows the conditions of tribal villages will not be surprised when she points out that of the nine important civic amenities not more than four were available in any of the 23 villages covered by the study. It is also not surprising that out of the 115 households 71 had encroached upon forest lands and that of these 10 households depended substantially and 82 households marginally on forest.

Four problems faced by the tribal households in using the forest produce, well listed by Dr. Ghate are: (1) lack of information regarding rights and concessions available to the forest dwellers, (2) exploitation by the forest officials, (3) distance, and (4) non-availability of forest produce. With increasing deforestation the problems are going to be more acute.

Dr. Ghate has discussed the nature of desirable long-term relationship of forest dwellers and forest. Community ownership and management may be ideal but appears impractical in the present socio-economic milieu. Forest Department cannot cope up with a problem of such magnitude. The only solution appears to be a combination of government ownership and, wherever possible, of joint management by the people and the Department. Experiments by Forest Protection Committees in West Bengal need to be scrutinised and replicated with suitable modifications according to local conditions. This calls for a drastic change in the attitude of forest officials and also of the forest dwelling communities.

I hope that Dr. Ghate's study will motivate other social scientists to undertake research into this grave problem of national magnitude and to work out suitable solutions in different regions in the country. The national crisis of accelerating deforestation which is also a part of global crisis can be resolved only by a determined and well-informed national effort.

SHARAD KULKARNI
Director
Centre for Tribal Conscientisation
Pune

ACKNOWLEDGMENTS

The present work is an attempt to study the changing economic condition of the tribals as a consequence of various forest policies adopted over a period of time. To get a first hand information I had to personally interact with tribals through a questionnaire. I must thank all of them for their cordiality, frankness and hospitability.

In the process of this research, I was fortunate to have personal talks, discussion and correspondence with some of the eminent scholars and workers associated with the problem of ecology and tribals. Shri S.D. Kulkarni needs a special mention as he provided me with volumes of literature on related topics and valuable suggestions. Thanks are also due to Shri Madhav Gadgil, Shri Ramchandra Guha, Dr. B.D. Sharma from whose works I have quoted extensively.

The Government's side of the problem is represented by the Forest Department of Maharashtra. A lot of information, suggestions, and guidelines were provided by various officials of the department. Preparation of the present work required an extensive amount of field-work to be taken in the interior parts of Maharashtra forests. The forest officials were very kind and hospitable and without their help, co-operation and guidance it would have been virtually impossible to visit the interior villages in forests and collect the information. To name a few I must thank Shri R.L. Choudhary, Shri R.N. Indurkar, Shri P.D. Ambaskar, Shri B.M. Parab, Shri G.M. Bedekar, Shri Ashok Sharma, Shri S.B. Elkunchawar, Shri R. Jadhav and Shri Waghmare.

I must thank the Head, Department of Economics, my colleagues and friends for their constant inspiration.

This work is an outcome of the problem suggested by Dr. S.V. Khandewale, Department of Economics, Nagpur University, Nagpur. No words can express my sense of gratitude to him for his constant inspiration and able guidance. I am thankful to the ICSSR for providing financial assistance in the form of publication grant-in-aid for this book.

RUCHA GHATE

CONTENTS

LIST OF TABLES

LIST OF FIGURES

PART I

The supply of bamboo, the main raw material for paper, is [illegible]. This has forced the producers to go in for unconventional raw materials like agricultural residues (rice and wheat straws, etc.), cotton rags, hemp, jute cuttings and grass, or mix hard-wood with bamboo. In such cases, both private and public sector paper mills are heading towards the last forest frontiers such as Bastar, the northeast and the Andamans.

The most important aspect of these concessions and subsidies is the treatment that the tribals got and the price which they had to pay for these materials. While the paper mills had to pay anything between 37 paise to Rs 22.80 for a tonne of bamboo, tribal artisans could purchase it in the open market only at Rs. 1,[illegible]20 per tonne. The Buruds, the traditional basket weavers, found it impossible to buy bamboo at the market rate and could not sell their costly product at enhanced prices. Tribals, who were getting their fixed quota of bamboo as their nistar rights, ceased to get it because of non-availability. In Bhandara district of Maharashtra, the tribals had to protest by organising a *rasta roko*, when they were denied their share of bamboo due to non-availability while truck-loads of bamboo were going to Ballarshah Paper Mill.

The depletion of bamboo forests in Tamil Nadu reduced the earnings of basket weavers from Rs. 5.00 per day in the early 1970s to less than Rs. 1.00 per day now, in 1979. Furthermore, presently these basket weavers are always under the threat of sudden periods of unemployment of several weeks, due to non-availability of bamboo, a situation they had never known in the earlier age.

In addition to its use as a raw material for basket weavers, [illegible] an important building material and it is also a [illegible]. Some tribals even store water or grains in big bamboos and use it as utensils. Such a useful article for tribals, bamboo, is becoming a [illegible] scarce because even now some state forest departments [illegible] bamboo forests to paper mills who leave behind nothing. While bamboo has disappeared from natural forest, the bamboo plantations in man-made forests do not recognise the rights of the tribals.

1

INTRODUCTION

The Problem

To a vast number of tribal people, the forest is a home, a livelihood, the very existence. It gives them food-fruits of all kinds, edible leaves, honey, nourishing roots, wild game and fish. It provides them with material to build their houses and things for practising their arts. By exploiting its produce they can supplement their meagre income.

From times immemorial the tribal people have enjoyed freedom to use the forest and hunt its animals, and this has given them a conviction, that remains even today deep in their hearts, that the forests belong to them.

This state of affairs continued happily until about the middle of the 19th century. Thereafter, people from outside began to move into the forests and conditions began to change. The gradual extension of the authority of the Government in these areas and the natural desire of the forest officials to exercise closer control over the use of forest products deeply disturbed the entire tribal economy, and sowed the seeds of conflict.

In recent years, there has been much talk about the rapid denudation of green cover of the geographical area of our land. Concern is expressed over the environmental pollution, ecological imbalance, soil erosion and the like, by various experts in various fields. Preservation and expansion of forest cover are advocated both as preventive and curative measures to eradicate all those ills.

Being aware of the situation, the Central Government and the State Governments of our country have substantially stepped up the spending on forest development. Government spending out of its own resources is supplemented substantially by foreign aid. Along with the stepping up of expenditures on forest development, reformulation of the present national forest policy has been discussed in the government and non-government circles. As a result of this a new forest policy incorporating some fundamental changes in outlook has evolved.

The long and persistent controversy centres round the apportioning of the blame for the existing grim situation of our forests between the forest managers and the forest dwelling community. Unless the causes of the present ills are properly diagnosed, no amount of money pumping can solve the problem.

The National Commission on Agriculture, in its report on forestry, submitted in 1976, had specially addressed itself to the problem of rapid recession of forest area in the country. The Commission has blamed the forest dwellers along for the disappearance of the forest. But it had not carefully examined the nature of their use of the forest products.

It is indeed very painful to observe that in spite of a general liberal commitment of the Constitution of India to protect the interests of the Scheduled Tribes and lead them on to a steady path of development, the national forest policy continues to reinforce the colonial policy of looking upon forests mainly as a source of state revenue and to restrict the traditional rights and privileges of the local communities. This general policy of 'forest officials and specialists *versus* people' continues unchanged in the Independent India. However, this outlook seems to have changed to some extent as is reflected in the recent Forest Policy of 1988.

Present Study

The inter-relationship between the tribals and the forests may be studied through various angles, such as income and employment generation through social forestry, evaluation of Tribal Welfare Schemes, impact of Forest Labour Cooperative Societies, cost-benefit analysis of Forest Development Corporation of Maharashtra, rehabilitation of project affected persons, working of tribal sub-plans, marketing of forest produce etc. In the present study, however, a wider perspective has been taken in establishing the economic impact of forest policy on the tribals in Maharashtra.

After the British devastated the Indian forests for their colonial purposes, the Indian Government have been persistently emphasising the need to bring back the level of forest cover to the required 33.33 per cent.

In our country, according to 1981 census, 7.76 per cent of the total population (9.19 per cent in case of Maharashtra) is tribal. More than 90 per cent of this tribal population lives in forests and depends on forests for its wherewithal. Uplift of the tribals is one of the declared objectives of the policy of the Government.

The state of affairs, however, is that so far neither the forest cover has reached the stated level of 33.33 per cent, nor has there been any improvement in the economic status of the tribals. It is felt that the present plight of the tribals is closely associated with the problem of gradual denudation of the forests and a peculiar perspective taken by the policy formulators.

The present work endeavours to examine the ongoing situation of economic relationship between the tribals, forest resources and the forest policy of the government. It further tries to pin-point the drawbacks of the forest policy *vis-a-vis* the present plight of the tribals. This study is an humble effort to examine the economic effects of the forest policy on the tribals. Looking at the plight of the tribals, the present researcher does confess to have a bias in favour of tribals. This bias has been converted into an assumption that though the tribals are illiterate and comparatively idle, there must be something seriously wrong with the policy-making at the governmental level and, therefore, at the implementation level, leading to their plight. With this underlying and implicit assumption, some other explicit assumptions have been made to build-up a frame-work for this study. In this dissertation the term tribal has not been used in anthropological sense. It covers not only tribals but refers to the local communities of forest dwellers also.

The objectives and the hypotheses of the study are as follows:

Objectives

1. To study the trend of changes in forest policy since the British period as incorporated in the Acts;
2. To study the impact of changes in the forest policy on the economic conditions of the tribals in Maharashtra;
3. To study and analyse the reactions of tribals to the changes in forest policy in Maharashtra; and
4. To envisage a long-term harmonious relationship between the forest policy and the tribals' economic interests.

Hypotheses

1. Forest policy formulation has neglected tribals' dependence on forests;
2. Standard of living of tribal communities has not improved in Maharashtra;
3. Forest administration and tribals seem to have an indifferent attitude towards each other; and
4. By educating and associating tribals in the management of forests, their standard of living can improve, and it can also check the denudation which is due to illicit cutting and encroachments.

Before describing the methodology adopted for this study, a brief review of literature available on the subject is presented.

Review of Literature

The subject of forests has been much discussed by ecologists in the past decade or so, mainly because of the large scale degradation faced all over the world. On the other hand, tribals were always fancied as a subject of study by anthropologists and sociologists. It is very recently that the academicians and social workers have brought the problem of tribals' rights and their dependence on forests to the fore. The review of literature presented, covers the main writings on this aspect of forests and tribals, and not the botanical details of forests or the social, religious or anthropological studies on tribals. Details, like the year, publication, source, publisher etc. are given in bibliography along with the list of many other publications on similar topics.

The Annual Report of Forest Department, published by the Government of Maharashtra covers the details on administration, silvi-cultural research, forest utilisation activities and economic research, financial position etc. The reports are published and are available from 1961 onwards i.e. after the formation of the State of Maharashtra.

The Report on 'The Scheduled Areas and Scheduled Tribes' by the Commission chaired by Shri U.N. Dhebur (1961) contains two main chapters on tribals and forest policy in which it has taken a brief review of British policy and has suggested some changes in the attitude of government.

Hundred Years of Indian Forestry (1961), brought out by the Forest Research Institute, Dehra Dun, has taken a review of the achievements made by Indian Forestry. The salient features of the British policy and various forest Acts, have also been enumerated. The economic problems of the tribals, however, have not been mentioned.

National Commission on Agriculture (Vol. IX) (1976) has dealt with the assessment of present low productivity of forests and has suggested measures to increase their productivity from the point of view of revenue. Tribals have been considered as obstacles, and curtailment of their rights has been advocated.

Concessions and Plans Related to Forests in the State of Maharashtra (1979-80), a publication by Tribal Research and Training Institute, Pune, has given detailed list of concessions in all the districts of Maharashtra.

Report of the *National Committee on the Development of Backward Areas* (1981) by the Planning Commission has limited itself to the review of the past and on-going programmes, and has made some suggestions regarding the role of the government in tribal welfare.

Tribal Research and Training Institute, Maharashtra State, Pune, has brought out a *Tribal Research Bulletin* (1981) which deals with some developmental aspects of the tribal areas. A case study of tribals of Gujarat is also presented.

Report on 'The Role of Women in Community Forestry' (1980), organised by the Ministry of Agriculture, covers the present status of women in forestry and their role in the collection of minor forest produce.

Report of the 'Committee for Afforestation of Barren, Denuded Lands and Hills of Maharashtra' (1982) contains field observations of the afforestation sites all over Maharashtra, and the results of the interviews of forest officials.

Various Plans for Tribal Development (1984) is the publication of Tribal Research and Training Institute, Pune. It contains the tribal development programmes undertaken by different departments like Agriculture, Forest, Home, Revenue, Animal Husbandry, Dairy, Horticulture, and Social Forestry.

Report of the seminar conducted by Marathi Vidnyan Parishad, Nagpur, on 'Role of Minor Forest Produce in Economic and Social

Development' (1986) mainly lists different minor forest produce available and their various uses.

Report of the Committee for Review of Rights and Concessions in the Forest Areas of India, (1984) chaired by Shri M.S. Chowdhary, IFS, Government of Maharashtra, has evaluated the damages caused by the concessions given to forest dwellers in Maharashtra. But this Report has not mentioned the problems of the forest-dwellers in availing of the concessions.

Lok Rajya the Adivasi Development Number, 1983, published by the Government of Maharashtra has articles dealing with tribal development and their problems, however only nominally covered.

Forestry in India's Economy, New Approach to Forest Management (1985) by Shri Ashok Sharma, IFS, gives details of Forest Development Corporation of Maharashtra — the aims and objectives and the achievements till 1985, published by Forest Development Corporation of Maharashtra.

S.D. Kulkarni, Director of Tribal Conscientisation Centre, Pune, has been writing on the tribal problems for several years. He has written about tribals' rights, their land alienation problem, encroachments etc., and has severally criticised government for its unjust bias against the tribals.

M. Gadgil from the Indian Institute of Science, Bangalore, in his several articles has emphasised that forest management by tribals is very much a desirable experiment, and has condemned the attitude of the Government of India towards tribals.

R. Guha in his articles criticises the colonial attitude of the government which has continued even after the Britishers have left. He has written against the draft forest policy of 1984 also.

Anil Agarwal has brought out *The State of India's Environment : A Citizen's Report* (1982 and 1985), which gives a very good picture of the use of natural resources like water, forests etc. The report also evaluates the present condition of dams, and projects like Social Forestry etc.

Desmand D'Abreo, in his book *People and Forests* has shown how tribals suffer because of sheer neglect of the policy-makers.

Erik Eckholam in his various articles deals very neatly with the tribal problems, and he has suggested measures to improve their economic condition along with forests.

G. Joshi in *Forests and Forest Policy in India* has taken a review of the past policy of the government and its adverse effects on the tribals.

K.P. Kannan is very critical about the Draft Bill of 1980 and has shown very clearly the methods adopted by the Government of India to increase their hold on forests and drive away the forest-dwellers, in number of his articles.

PUDR, too, has published its severe criticism against the Draft Forest Bill of 1980 in *Unclassed Civil War : A Critique of Forest Bill*, where it has quoted extensively from the Bill, showing increasing power of the government in forests.

B.D. Sharma, in *Forests, Tribal Economy and Regional Development* has shown the evolution of the man's relationship with forests, and has argued that forests can be developed along with tribals only.

Walter Fernandes and Sharad Kulkarni have edited a book titled *Towards a New Forest Policy* (1983) which comprises several articles by scholars concerned with people's rights, and environmental needs of forests.

B.K. Roy Burman is a known scholar in the field of forests and tribals and in the paper, *Issues in Environmental Management Centering Forest and Role of Tribal Communities*, he has emphasised the importance of teaching in tribal communities for rural development.

Dr. Chowdhary in her article *Managing Forests for Development*, has taken historical perspective of the British management and states that new institutions are needed to manage the forests.

Col. B.L. Verma in *Vanishing Forests in Maharashtra* has shown the land utilisation pattern, and has proved that hydel projects are white elephants.

Dr. S.G. Deogaonkar, in *Problems of Development of Tribal Area*, deals with the socio-cultural and economic problem of tribal development.

L. Fortman in *Land Tenure, Tree Tenure and the Design of Agro-Forestry Projects* has suggested that forest dwellers should get the profits of tree plantation and that this alone can assure the safety of plantation programmes.

V.J. Patel has suggested methods of meeting fuelwood crisis of today and ensuring the supply for tomorrow in his article 'Rational Approach Towards Fuelwood Crisis in Rural India'.

Sunderlal Bahuguna, the father of the 'Chipko Movement', very aptly shows how the people who are blamed for felling are actually saving forests while the government is responsible for the general degradation.

'An End to Laughter, Tribal Peoples and Economic Development' (1985) a study by Survival International Review has presented disturbing evidence which shows that instead of benefiting from economic development, tribal peoples are often its principal victims. This timely report also warns of a major threat facing over two millions tribal people in India, who are likely to join the great mass of landless poor, if plans for a series of vast, all ill conceived hydro power projects go ahead.

N.G. Basu in *Forests and Tribals* (1987) has examined the aspects of community forestry and the role of tribals in it. The book also presents village studies conducted in West Bengal.

'Our Common Future' (1987), a Report published by The World Commission on Environment and Development has re-examined the critical environment and development problems on the planet. It has formulated realistic proposals to solve the problems and to ensure that human progress will be sustained through development without bankrupting the resources of future generations.

'The Tropical Forestry : Action Plan' has been prepared by FAO in cooperation with the World Resources Institute, the World Bank and the United Nations Development Programme. The plan has five components : forestry in landuse, forest-based industrial development, fuelwood and energy, conservation of tropical forest eco-system, and institutions to reach the people.

Vandana Shiva in her *Forestry Crisis and Forestry Myths : A Critical Review of Tropical Forests : A Call for action* has made a critical evaluation of the US $8 billion plan to save the world's tropical forests. She points out that the plan in reality will cause serious negative impact on the forest ecology and the people who depend on it. The report further says that the plan's major emphasis on industrial forestry will only benefit the rich First World countries whilst leading to the further improverishment of the Third World Communities and tribal peoples, and the wholesale destruction of the natural tropical forests.

Methodology

The dissertation is divided into three parts. The first part consists of the selection of the problem, review of various forest acts and the economic impact of forest policy on tribals in general. The second part contains description of forests in Maharashtra, selection of samples for field study and results of the field study. In the third part contours of long term relationship between forests and tribals have been envisaged, and a summary of conclusions, and suggestions have been presented.

The first chapter of the first part gives the introduction to the subject. It has also given the back-ground of the problem and the methodology used for studying the problem. In the second chapter, a review of the changes in Government's forest policy from the advent of British has been undertaken. The underlying motive of the government to use forests commercially and fetch maximum revenue even at the cost of the forest dwellers, has been pin-pointed. The third chapter deals with the economic impact of this policy on the tribals affecting them directly and indirectly. This portion of the study is based on secondary data collected from various government reports, studies of scholars and statistics published by various departments and agencies.

The second part of this dissertation consists of fourth and fifth chapters. In the fourth chapter brief description of forest and tribal areas, tribal population of Maharashtra has been given. Then the selection of three sample used for field study, and the methodology used for the field study have been explained. The three samples used are as follows:

The first sample is a multi-stage sample which consists of randomly selected (115) households from purposively selected 23 forest villages from four forest circles of Maharashtra.

The second sample consists of interviews of ten forest officials ranging from the rank of Range Forest Officer to Conservator of Forests or equivalent, either from the Forest Department, Forest Development Corporation of Maharashtra or the Department of Social Forestry of Maharashtra.

The third sample consists of a group of 'typical villages', all from Maharashtra. They are:

(a) an unauthorised settlement;
(b) an expropprietary forest village;
(c) a village settled by the Tribal Development Corporation of Maharashtra;

(d) a Kathiawadi settlement; and

(e) an irrigation-project settlement created by the Revenue Department.

Fifth chapter gives the results of the field study based on the observations made from the interviews with the help of pre-tested question schedules.

The third part of the dissertation consists of two chapters. Sixth chapter envisages a situation of simultaneous development of forests and tribals. The last chapter incorporates the summary of conclusions and suggestions based on the study undertaken.

2

VARIOUS FOREST ACTS

In this chapter, it is proposed to take a review of the policy adopted by the British as well as the Indian Government regarding forests and tribals. Forests, as a natural resource, are useful to all. But they have a special significance for those who dwell in and around forests since ages. How this aspect was deliberately neglected by the policy formulators, is the subject matter of this chapter.

Gravity of the Forest Question

The subject of people's rights in forests is complex and controversial and it has been attracting a lot of attention over the years.

In Prussia, the gathering of dead-wood had traditionally been unrestricted, but the growing needs of industry led to legal controls on the free access to the forests for the forest dwellers. Marx argued that the State should defend customary law against the rapacity of the rich. Marx maintained that the restrictions were the result of advice given by rich class to law givers, ". . . that, when making a law about wood and timber, they are to think only of wood and timber, and are not to try to solve the material problem with the whole complex of civic reasoning and civic morality".[1]

Similar was the situation in India. Before the advent of the British, forest dwelling communities had free access in forests to collect any forest-produce that they required. During the British rule, commercial use of forests increased manifold, which resulted in more and more restrictions being imposed on the tribals residing in forests. And this attitude continued even after Independence.

This chapter is divided in three sections. The first section deals with the situation that prevailed in the country before the Britishers came into the picture.

In section II, there is a detailed discussion of the British forest-policy and the Forest Acts, which were framed between 1855 and 1947.

Section III deals with the forest policy adopted by the Independent India towards the forests and the tribals.

SECTION I

PRE-BRITISH PERIOD

"The story of man can be narrated in terms of his changing relationship with nature and forests",[2] so said the then tribal Commissioner of Madhya Pradesh, Dr. B.D. Sharma, in his paper on 'Forests, Tribal Economy and Regional Development'. Man appeared in the world as a part of the 'nature system' depending entirely on natural produce. He developed from 'animal-like state' into a hunter, a food gatherer, and a cultivator, but remained an integral part of the ecological system.

This section deals with the type of forest management that prevailed in India before the Britishers came to rule.

Forest Management

Though very little is known about the management of forests before the advent of the British, a perusal of the travelogues of the early 19th century and the Gazetteers gives us a picture of a well-wooded country whose pasture and forest resources were controlled and fairly managed by the local village communities. The forests were a communal property with no individual staking any claim to them though all the individuals had recourse to forests for their needs. For the tribals in particular, the forests remained a life-blood. The symbiotic relationship of the tribals with the forests constituted the foundation of the independence of the tribals from the outside world.

In the initial periods of the history of Indian Forestry, stretching roughly from the beginning of the settled agriculture right up to the middle decades of the last century, State control over forests was virtually non-existent. In the periods when the military technology of large empires relied heavily on forests for timbers and elephants, the State made explicit provision for forests to be set aside for the use of the subject population. As a result India's vast forests were largely under the control of village and tribal communities.

Rational Use

The people in various forest areas of India, developed a variety of social and cultural institutions which enabled prudent utilisation of forest produce, such as :

(a) Restrictions on indiscriminate exploitation of resources so that a sustained supply is assured;

(b) Restrictions on exploitation to only certain stages of life: the hunter-gatherer tribe of 'phasepardhis' of Maharashtra traditionally do not kill fawns and pregnant pigeons, or black-buck even if snared;

(c) Protection to certain 'sacred' species : the *Banyan* and *Peepal* are traditionally protected throughout India, while *Khejdi*, a locally very useful tree, is protected in the Bishnoi villages of Rajasthan;

(d) Protection to entire ecosystem : certain patches of forests or ponds or stretches of rivers are offered total protection. For instance, on the Kerala-coast, there are sacred serpent groves which preserve the rare specimens of certain plant species in India;

(e) Limited exploitation of certain resources : In many tribal village forests, there is a firm social control over the quantum of exploitation by each family. For instance, only one headload of fuel per week per family may be permitted to be taken out;[3]

(f) The Pnars of Meghalaya preserve their sacred bush as a religious duty; and

(g) Among the Angamis, when certain species of plants are extracted, custom requires double the number to be planted.[4]

Though these were only customary regulations of people's rights over forest-lands and forest-produce, it did not pose any problem due to the existence of vast tracts of forests and a relatively small population of tribals. The communities, though inequitous in many ways, had mutual trust and common interests so far as the forests were concerned.

We know very little as to how these traditions of prudence came to be accepted and practised all over India. However, they illustrate every type of prudent management prescribed in the modern times. They must have gradually come to be adopted on the basis of experience and common sense of the people.

Community Ownership

The Government of India Committee on Forests and Tribals (1982), after studying the situation on the spot, commented that ". . . we cannot help referring to our experience in the North-East where we found a number of villages maintaining their communally owned forests which were very well managed".[5] Community ownership was a common feature in India, which was also the reason for superb preservation. The welfare of the community being superior to any individual, forests were considered a perpetual source of food, fodder and fuel for the forest dwellers.

In brief, those who lived in the forests also looked after the natural resource well. The travelogues of pre-British days give a sufficient proof of this. Forests being a communal property, whole community felt itself responsible towards its preservation. It is only when the British came that the things changed drastically.

SECTION II

THE BRITISH PERIOD 1855 – 1947

With the Britishers, came the 'scientific management' of forests which served their own purpose well. Tribals became a hindrance in the exploitation of forests on which the British defence relied heavily. Therefore, every successive policy and legislation formulated and adopted by the British Government, resulted in increased restrictions on tribals. A review of the British policy is undertaken in this section.

Conservation Neglected

The early days of the British rule, up to the middle of 19th century, were characterised by a total indifference to the needs of forest conservancy. Forests were considered obstructions to the prosperity of the country and their removal meant increase in the land under cultivation and hence increase in the revenue. The whole policy of the time was to extend agriculture, and the watchword was to destroy the forests with this end in view. A corollary of this policy was that the people's rights in forests continued to be exercised unrestrictedly.

The First Reservation

By the 16th century, much of the Western Europe had been deforested, and Northern Europe was supplying Britain with large quantities of wood. Between 1600 and 1700 A.D., Ireland's forests were devastated to meet England's needs for timbers for ship building, iron smelting and tanning. With oak forests vanishing in England, a permanent supply of suitable timber was required for Royal Navy as the safety of the empire depended on its wooden walls. This was the period of fierce competition between the colonial powers, and Indian teak — the most durable of ship-building timbers, which saved England during the war with Napoleon. The show of interest — the reservation of teak in Malabar in 1806 — was predictably dictated by '. . . imperialistic considerations for revenue and timber for defence'.[6]

Railway Expansion

On consolidating their hold over India, the British began to lay down an extensive network of railway lines to facilitate the export of raw materials from and the import of finished goods to the remote corners of the country. "Often these lines were deliberately routed through the richest areas, as for instance the diversion via Londa between Dharwar and Belgaum in Karnataka, to tap the forest resources".[7]

The early years of railway expansion saw an unprecedented assault on the easily accessible forests. Great chunks of forests were destroyed to meet the demand for railway sleepers, no supervision was done over the felling operations, and a large number of trees were felled whose logs could not be utilised. This depredation finally awoke the colonial authorities to the fact that India's forests were not inexhaustible.

Establishment of Forest Department

In 1862, the Governor-General called for the establishment of a department that could ensure the sustained availability of the enormous requirements of the different railways for sleepers. As an officer later reminisced, the forest department was started because the government became aware that the magnificent forests of India and Myanmar were being worked by private enterprise ". . . in a reckless and wasteful manner and were likely to become exhausted if supervision was not exercised".[8] The criticism was all the more acute since only three Indian timbers — teak, *sal* and *deodar* — were strong enough in their natural state to be utilised as railway sleepers.

It was in this situation that the Imperial Forest Department was formed in 1864, with the help of German foresters.

Tribals became "Intruders"

In response to the above-mentioned situation, the guardians of forests (the Imperial Crown), declared the people of the forests as intruders, aliens over the 'state property'. However, tribals were allowed to live in forests because they provided the cheap labour by which the forests could be cut, the timber transported and readied for the timber trading. The tribals thus ". . . formed an appendage to the colonial interests".[9]

The government felt that for effective functioning of the department, the previously exercised unlimited rights of the users should be restricted. All this was done in the name of conservation of forests. It is very strange that the general interests of the public came to be enshrined while those of the tribals came to be restricted.

A memorandum providing guidelines restricting the rights of the forest-dwellers (for conservation of the forests) was issued in August 1855 and was later modified in 1884. It proclaimed public benefit and added that, ". . . in almost all cases the constitution and preservation of a forest involves, in greater or lesser degree, the regulation of rights and the restriction of privileges of the users in the forest-area". It was also mentioned that the reservation was done for greater public benefit and the cardinal principle to be observed was ". . . restrictions only in such degree as is absolutely necessary to secure advantage".[10]

This last point was always overlooked and successive forest legislations put more and more restrictions on people's rights over forests. The first attempt at asserting State monopoly right was made through 'The Government Forests Act, 1865'.

THE GOVERNMENT FORESTS ACT, 1865

Increased Powers of the Government

The Act empowered the government to declare any land covered with trees, brushwood or jungle as government forest by notification, provided that, ". . . such notification should not affect any existing rights of individual or communities".[11] On the other hand, the governments could also prescribe punishments for the breach of the provision of the Act and rules were laid down for confiscation of implements used in the offences and for the arrest of offenders. It can be seen from the provisions of the Act that the major objective was to establish government's control over forests because by the definition of 'offence', tribals could do nothing in forests — not even collect fallen leaves.

Restrictions on Tribals

"Rules made in pursuance of this Act provided for the following matters — The preservation of all growing trees, shrubs and plants, within government forests or of certain kind only — by prohibiting the marking, girdling, felling and lopping thereof, and all kinds of injury thereto; by prohibiting the kindling of fires so as to endanger such trees, shrubs, and plants; by prohibiting the collecting and removing of leaves, fruits, grass, wood-oil, resin, wax, honey, stones, lime or any natural produce of such forests; by prohibiting the ingress into and the passage through such forests, except on authorized roads and paths; by prohibiting cultivation and the burning of lime and charcoal, and the grazing of cattle within such forests."[12]

It is instructive to note that it was the Britishers who recklessly destroyed the Indian forests. But when it came to imposing of restrictions, the tribal people were the victims. Restrictions being imposed on all the basic activities, the forest-dwellers, once the owners of forests, became the puppets in the hands of British rulers.

INDIAN FOREST ACT, 1878

New Definitions

At this stage, the British government felt the necessity of settling the claims of tribals on land which was declared as forest land. The Indian Forest Act, 1878, was, therefore, more comprehensive than the earlier one and defined several terms precisely and for the first time:

'Tree' included palms, bamboo, stumps, brushwood and canes;
'Timber' included trees when they have fallen or have been felled, and all wood, whether cut up or fashioned or hollowed out for any purpose or not.
'Forest produce' included —

(a) the following, whether found in, or brought from, a forest or not, that is to say — timber, charcoal, caoutchouc, catechu, wood oil, resin, natural varnish, bark, lac, *mahua* flowers, and myrabolams; and
(b) the following when found in, or brought from, a forest, that is to say:

  (i) trees and leaves, flowers and fruits and all other parts or produce not hereinbefore mentioned of trees;
  (ii) plants not being trees (including grass, creepers, reeds and moss), and all parts or produce of such plants;
  (iii) wild animals and skins, tusks, horns, bones, silk cacoons, honey and wase, and all other parts or produce of animals; and
  (iv) peat, surface soil, rock, and minerals (including limestone, laterite, mineral oils and all products of mines or quarries).[13] Similarly 'cattle' and 'river' were also defined in details.

Classification of Forests

For the first time, in this Act, forests were classified as —

1. Reserved forests,
2. Protected forests, and
3. Village forests.

Government had powers to reserve any forest according to the provision of the Act. "The Local Government may from time to time, constitute any forest land or waste land which the property of Government, or over which the Government has proprietary rights, or to the whole or any part of the forest produce for which the Government is entitled".[14]

Reserved forests can be briefly defined as — forests, where

everything is restricted unless permitted. Protected Forests are — forests, where everything is permitted unless restricted.

The Government could declare any class of trees in a protected forest or any tree in any such forest, to be reserved from a date fixed by notification, declare a portion of such forest to be closed or prohibit removal of any forest-produce. The forest produce, timber or grass, was allowed to be had against a licence granted to the inhabitants of the towns and villages in the vicinity of protected forests.

Regarding village-forests as well, the Government retained the authority to cancel or assign the rights to village community though these forests were to provide timber, other forest produce and pasture to the people.

Forest Settlements

Persons were to be notified to prefer their claims over land and forest produce in any proposed reserved forests. Forest settlement officers were appointed to record such rights and to make some special provision to ensure exercise of such rights. Every person had to appear before the officer, within a fixed period, to register, ". . . the nature of right, the amount and particulars of the compensation claimed in respect thereof".[15]

The forest settlement officer was empowered to enter, by himself or any officer authorised by him for the purpose, upon any land, and to survey, demarcate and make a map of the same, stop any public or private way or water course in a reserved forest. In the trial of suits, he was given powers of a Civil Court. He was also empowered, in the case of a claim to rights of pasture or to forest produce, to pass an order admitting or rejecting the same in whole or in part.

At the most fundamental level, the demarcation and fencing of large tracts of reserved forests meant effective loss of control by the forest dwellers over their habitat — a control so necessary for the reproduction of their existence. This also, inevitably, increased the pressure of tribals, dependence on village forests, thereby hastening their denundation.

Tribals turned into Cheap Labourers

There was one positive aspect of the industrial revolution in Europe that the whole population gradually came to share the benefits of the more intensive resource use. Unfortunately for India, effect of the change in Europe was entirely negative. The bulk of tribal population not only did not come to share the benefits of more intensive resource-use, but was even denied the traditional low level of usage which was essential for their day to day survival.

The wages the labour force received for working in reserved forests were paltry. The contractors, even in those days, preferred to bring outside labour instead of using the local labour supply. The forest revenue thus generated flowed straight into the state coffers, with no share for the local governmental bodies such as panchayats. It was rightly criticised that, ". . . to the rural population then, land under reserved forests was land alienated from them".[16]

Derecognition of Communal Ownership

In the pre-British period, a certain portion of the country's land was set aside for communal use by the villagers. But the British Government made no attempt to ensure that these lands were well managed, or their resource augmented by planting of useful species. Dr. Voelekar's report on 'Indian Agriculture' noted in 1885 that the productivity of Indian agriculture was falling because farmers were being increasingly forced to use dung as fuel. He urged the forest departments to give up their single minded pursuit of timber and devote serious attention to the creation and good management of fuel and fodder plantation for the rural population. This advice was not heeded and the forests were increasingly impoverished. The result was that, ". . . the disorganised villagers continued to abuse the village commonlands and suffer further shortage of their essential requirements".[17]

Derecognition of communal control of property was one of the instruments government wielded to achieve its goal. In the pre-British period the Indian villages used to pay revenue to the rulers as a community rather than individually. The British insisted on individual land settlement and payment of revenue at an enhanced rate.

Another important change was that the government also tried to claim as State property all the village common grazing lands and forests. Though the State had, in certain areas, made over some forest under settlement for the villagers to utilise as village-forests, the loss of community ownership had effectively broken the link between the man and the forest. The harmony of village communities was rudely shaken with derecognition of communal rights and introduction of money economy.

Restrictions by Princely States

The large areas of forests under the Princely States were also drawn in, though indirectly, into the orbit of colonial capitalist expansion. The exploitation of these forests, either by agency of the colonial state, or in the later stages, directly by the Rajahs was instrumental in introducing a qualitative change in the relationship between the rulers and the subjects.

Over the time, the native rulers, who were very indifferent towards their large forest wealth, became, ". . . very much alive to the value of their forest properties as they severely curtailed the traditional rights of the subject communities".[18]

National Forest Policy, 1894

This policy was primarily based on the recommendations contained in the report of Voelekar (1893) on Indian Agriculture, governed by the British needs of agriculture resources for industries from 1860 onwards. The policy recognised the close relationship between forestry and agriculture and argued that the justification for forestry activities springs from their direct and indirect contribution to the development of agriculture. Therefore, even if certain areas were found to be suitable for growing good quality timber, they were not utilised for forestry. Such an approach facilitated the extension of agriculture to forest areas, ". . . largely benefiting the landed interests and adversely affecting the adivasis for whom the concept of property rights over land was almost unknown".[19]

Aims of the Policy

The British forest policy embodied, in brief, these aims:

(i) The sole object with which State forests are administered is public benefit. In general the constitution and preservation of the forest involves the regulation of rights and the restrictions of the privileges of the user in the forest by the neighbouring population;

(ii) Forests situated on hill slopes should be maintained as protected forests to preserve the climate and physical conditions of them from the devastating action of hill torrents;

(iii) Forests which are reservoirs of valuable timber should be managed on commercial lines as a source of revenue to the State;

(iv) Wherever the effective demand for cultivable land exists and can only be supplied by forest area, the land should ordinarily be relinquished without hesitation, subject to the following conditions:

(a) honeycombing of a valuable forest by patches of cultivation should not be allowed; and

(b) cultivation must be permanent and must not be allowed to an extent as to encroach upon the minimum area of forest that is needed to meet the reasonable forest requirements, present and prospective.

(v) Forests that yield only inferior timber, fuelwood or fodder or are used for grazing should be managed mainly in the interest of the local population, care being taken to see that the use is not exercised so as to annihilate its subject and the people are protected against their own improvidence.[20]

Types of Forests

The forests were broadly classed under the following headings in the policy:

(a) Forests, the preservation of which is essential on climatic or physical grounds;

(b) Forests, which afford a supply of valuable timber for commercial purposes;

(c) Minor forests; and

(d) Pasture lands.

The first type of forest was, ". . . beyond any comparison with any other interests,"[21] even second type of forests which included great tracts from which supply of valuable timber like teak, sal, deodar and the like was obtained were, "... encumbered by very limited rights of user and mainly should be managed on commercial lines as valuable properties".[22]

Regarding cultivation, the policy observed that, ". . . the cultivation in question must not be allowed to extend as to encroach upon the minimum area of forest management".

Dr. I.H. Hutton, a distinguished member of the I.C.S., who made a deep study of the problems of the tribal people, wrote as follows about the effects of British rule on tribals — "Far from being of immediate benefit to the primitive tribes, the establishment of British rule in India did most of them much harm than good. It may be said that the early days of British administration did very great detriment to the economic position of tribes through ignorance and neglect of their rights and customs . . . many changes have been caused incidently by the penetration of the tribal country, the opening up of communications, the protection of forests and the establishment of schools, to say nothing of the openings given in this way to christian missions. Many of the results of these changes have caused acute discomfort to the tribes".[23]

Commenting on this Prof. G.S. Ghurye said that, ". . . the acuteness of discomfort was very often so great that it led to apathy, indifference, moral deterioration and even a decline in population".[24]

It is clear that this policy gave maximum importance to the revenue aspect of forests and commercialised supply of major forest produce needed in order to comply with the general forest needs of the country.

The third type of forests included tracts which produced only inferior sort of timber, yet, were managed as commercial undertakings, as they supplied fuel for manufacturers, railways and similar purposes in some cases. The policy also added that, "... it is not the intention of the Government of India to forego all revenue from the large areas that are valuable chiefly for the fuel and fodder which they yield".

Even the pasture lands were not to be totally under community ownership. The policy states that 'the fourth class of forests referred to as pastures and grazing grounds proper, which are usually forests only in name, can be conveniently declared as forests in order to obtain statutory settlement of rights,' hence by indirectly taking them away from the tribal ownership.

It is well known that the British had totally deforested the Eastern United States and South Africa during the 18th and 19th centuries. Therefore, it is quite wrong to believe that Indian forests were ruthlessly destroyed by its own inhabitants, instead it were the British who did so through the guise of rational and scientific forestry.

INDIAN FOREST ACT, 1927

The Indian Forest Act of 1878 was modified in parts by different acts of the local governments. It was later on replaced by the comprehensive Indian Forest Act, 1927. This piece of legislation contained all the major provisions of the earlier policy.

While all the restrictions proposed earlier remained, the penalty for defaulters increased. Punishment for the 'acts in contravention of notification' in the Protected forests, was increased from imprisonment for six months or fine up to five hundred rupees, or both, to imprisonment for Forest Officers or Police Officers who vexatiously and unnecessarily seize any property on pretence of seizing illegal property, remained the same i.e. imprisonment for six months or fine up to five hundred rupees or both.

Except such minor exceptions, the Indian Forest Act, 1927, was, word to word, the same as the Indian Forest Act, 1878.

The Government of India Act, 1935, passed by the British Parliament created provincial legislature. The subject of forests was included in the provincial list under this Act (item 22). Thereafter several provinces made their own laws to regulate forests. Most of these laws were within the framework laid down in the Forest Act, 1927.

Essence of the British Policy

Britishers formulated and reformulated their forest policy with the only motive of revenue collection. Forest department was established to improve authority over the tribals who were converted from owners of forests to cheap labourers. Increasing powers of government meant increasing restrictions on the tribals. This is the period when, for the first time, a formal forest policy came into existence, in the form of forest Acts.

SECTION III

POST-INDEPENDENCE PERIOD (1947-1984)

After Independence, all — tribals, social-workers working amongst tribals, academic scholars — hoped for some fundamental change in the very basis of forest policy. The government was expected to recognise the relationship of tribals with forests and get rid of the exploitation of tribals which was earlier practised in the name of 'forest management'. The government of free India disappointed everybody by adopting all the basic principles laid down by the Britishers. Tribals remained intruders and labourers, and did not get back their erstwhile rights over forests and forest produce. The present section deals, in details, with the steps taken by the Government of India.

Rethinking by the Government

After Independence there was some rethinking on the issue of forest policy which emphasised ecological and social aspects of forestry and gave only secondary importance to the needs of commerce and industry as also to the needs for revenue. But, somehow, it did not accept the contention that the neighbouring areas are entitled to a prior claim over forest and its produce.

The concept of national interest was interpreted in a very narrow sense. The destruction of forests for the construction of roads, building up of irrigation and hydroelectricity projects, ammunition factories and other projects was justified in the name of national interest, whereas ". . . cultivation on lands by the poor shown as forest lands without any actual tree cover, was treated as encroachment".[25]

THE NATIONAL FOREST POLICY, 1952

The continuity between the colonial and the post-colonial forest policy is provided by the National Forest Policy of 1952. The policy statement of 1952 affirms that the 1894 policy constitutes the basis for the forest policy of India up to this date because its " . . . fundamental concepts still hold good".[26]

Basis of the Policy

The policy of 1952 was formulated on the basis of six paramount needs of the country, namely —

(1) the need for evolving a system of balanced and complementary land-use, under which each type of land is allotted to that form of use under which it would produce most and deteriorate least;

(2) the need for checking —

   (a) denudation of mountainous regions, on which depends the perennial water supply of the river system whose basins constitute the fertile core of the country.

   (b) the 'erosion progressing space' along the treeless banks of the great rivers leading to ravine formation, and on vast stretches of undulating wastelands depriving the adjoining fields of their fertility;

   (c) the invasion of sea-sands on coastal tracts, and the shifting of sand dunes, more particularly in the Rajputana deserts;

(3) the need for establishing tree lands, wherever possible, for the amelioration of physical and climatic conditions, promoting the general well-being of the people;

(4) the need for ensuring progressively increasing supplies of grazing, small wood for agricultural implements, and in particular for firewood, to release the cattle dung for manure to step up food production;

(5) the need for sustained supply of timber and other forest produce required for defence, communications and industry; and

(6) the need for the realisation of the maximum annual revenue in perpetuity consistent with the fulfilment of the needs enumerated above.[27]

Classification of Forests

To fulfil the above-mentioned functions, the classification of forests was suggested as —

(a) Protection forests, i.e., those forests which must be preserved or created for protection of physical and climatic conditions;
(b) National forests, i.e., those which have to be maintained and managed to meet the needs of defence, communications, industry, and other general purposes of public importance;
(c) Village forests, i.e., those which have to be maintained to provide firewood in order to release cow-dung for manure, and to yield small timber for agricultural implements and other forest produce for local requirements, and to provide grazing for cattle; and
(d) Tree-lands, i.e., those areas which though outside the scope of the ordinary forest management, are essential for the amelioration of the physical conditions of the country.

It is obvious that the needs of tribals or forest-dwellers have been given low weightage or even that they are conspicuous by the absence of their mention. Protection forests are kept out of reach of tribals for ecological purposes, and so are the National forests as " . . . they constitute valuable timber bearing regions, the produce of which is indispensable for defence, communications and vital industries".[28] Their management on scientific and business lines was felt to be essential for maintaining a sustained supply of wood for industry and of large timbers for defence, communications and other national purposes.

'Narrow' National Interests

The National Forest Policy of 1952 observed that if the following two considerations are given undue weight, it would result in 'destruction of national well-being'. They are :

(1) neighbouring areas are entitled to a prior claim over a forest and its produce;
(2) agricultural requirements have a preferential claim over forest lands.

This attitude restricted the tribal claims on the one hand and did not give more importance to cultivation at the cost of forests, on the other.

The fact that the city dwellers and the city borns naturally get larger advantage of urban facilities of which forest dwellers are permanently deprived, was completely neglected by the policy of the Independent India also. Actually by the same logic, forest dwellers should get relatively more benefit. But, strangely, the policy added, specifically, that the " . . . claims of neighbouring communities, should in no event be permitted at the cost of national interest. The accident of village being situated close to a forest does not prejudice the right of the country as a whole to receive the benefit of a national asset", as if tribals are not a part of the whole nation.

The policy does not suggest anything for improving the quality of cattle that tribals own, but criticises to the extent that, " . . . the notion that a farmer's wealth must be reckoned in terms of the number of cattle he owns, regardless of quality is one of the causes of India's uneconomical cattle wealth".[29]

Industries Favoured

By 1952, wood-based industries such as pulp, paper and plywood were well established, and their demand for raw material became an important consideration in the management of the forests. The British had deliberately thwarted the growth of Indian industry till the First World War. After the war, they felt sufficiently secure of their own position to permit some industrialisation in their colonies.

On achieving Independence, our nation's preoccupation was to industrialise and bring the benefits of science and technology to the society. This is considerably justifiable, " . . . for modern technology has much to offer to man by way of a better life for the people, though urban people"![30] And paper, so essential to the spread of mass education in a democracy, is one example of an industrial product that needed production of bamboo etc. on a large scale. While all the attention concentrated on urban development, the rural, the tribal population was grossly neglected.

THE NATIONAL COMMISSION ON AGRICULTURE, 1976

The Commission advocated commercialisation of agriculture at all costs with almost a total disregard for the sustenance of tribals derived from the forests. Removal of poverty attainment of self-reliance and maximisation of employment were declared to be the objectives of planned development. At the same time it was laid down that, " . . . production of industrial wood has to be the *raison d'tre* for the existence of forests",[31] as if forests did not exist before the need for industrial wood arose.

Bias Against Tribals

The commercial and anti-poor bias of the Commission is evident everywhere in the Report. The Commission does not make it clear as to who has over-exploited the forests and simply presumes that the tribals have done it. Neglecting the fact that the supply of forest raw material to industries has been heavily subsidised and, in a relative sense, is 'dirt-cheap' or 'almost-free', the Commission comes heavily on the concessions given to the forest-dwellers, holding them responsible for the destruction of forests. It opines that, "Having over-exploited the resources, tribals cannot in all fairness expect that somebody else will take the trouble of providing them with forest produce free of charge".[32]

This was contradictory to what Dhebur Commission had observed on the privileges of tribals. Said the Commission: "We were rather surprised to find that there were no facts to substantiate the consistent propaganda against the tribals that they were asking for too much. . . . Against the estimated forest revenue of about Rs. 5 crores in Maharashtra in the year 1960-61, the value of forest produce removed by privilege holders and free grantees was only Rs. 5 lakhs in 1959-60".[33] It works out to mere one per cent. (This point has been discussed in Chapter 4 in details).

The National Commission on Agriculture not only holds tribals responsible for depletion of forest wealth but also finds them to be a real hurdle in the development of forests, by saying that, "The rights and concessions have also been convenient tools for entering the forests for carrying out clandestine removal of forest produce".[34] Hence it recommends that 'While the needs of the local population must be met to a reasonable extent, national interests should not be sacrificed', and also that, "no occupancy or permanent rights should be conferred upon the forest villages settled in forests". While the Commission comes down heavily on tribals holding them responsible for illicit felling, some of the press reports give a different story. According to a local daily, unhindered illegal trade in forests wealth like bamboo, charcoal, timber and firewood etc. is being carried over the past four years which has denuded the once rich forest in the Maharashtra-Madhya Pradesh border between Bortalao and

Darekasa. It further alleges, "In this illegal trade, Maharashtra and Madhya Pradesh have been deprived of crores of rupees worth forest wealth by contractors and merchants in collusion with forest and Railway officials who had reportedly received favours amounting to lakhs".[35]

While the Commission comes severely on the tribals, it is very mild on subsidies given to industries, saying, ". . . in the name of encouragements to industries, forest material has been leased out for longer periods at nominal rate of royalty. The present supply of forest-materials to forest based industries is highly subsidised. If forests are bled both by the rural and industrial sector, there will be no incentive on initiative left to change over from the low-cost, low-yield forestry to commercial high-investment economic forestry development".[36]

Basis of New Policy

The Commission recommended that the revised national forest policy of India should be based on important needs of the country. The needs were listed as —

1. Managing the forest resources of the country so as to provide maximum goods and services for the well-being of the people and economic progress of the country.
2. Checking denudation and erosion in mountainous regions and catchment areas of rivers on which depend perennial stream flows, fertility of the land in the catchment area and the useful life of dams and reservoirs.
3. Preventing erosion along treeless banks of rivers, and on the vast stretches of wastelands, and arresting the spread of sea-sands on coastal tracts and of shifting sand dunes in western deserts.
4. Maximum forest productivity with a view to meet the growing demand for industrial raw materials, timber and other forest produce for defence, communications and other domestic needs, and to augment the development potential with the ultimate aim of national prosperity.
5. Providing grass and grazing for livestock in forest areas, ensuring that it is not harmful to forests.

6. Providing small timber and fuelwood requirements of the rural population.
7. Providing recreational and tourist opportunities in the forests without impairment of forest resources including wild life and the preservation of environmental balance.
8. Creating blocks of forests interspersed with cultivation or by introducing trees in larger number in the ecosystem after careful selection for maintaining a healthy relationship between soil, vegetation and animal life and establishing diverse biotic complexes for minimising chances of elimination of enemies of insects and other pests which damage mono-culture.

It is clear that in the national policy formulation the environmental considerations and individual needs are given greater weightage than the basic needs of forest dwellers to the extent that there is a provision even for outside tourists but there is no mention of the tribal welfare.

Classification of Forests

The bias is proved further by the Commission's functional classification of forests, where the new concept of 'production forests' is introduced and emphasised. The classification suggested is as below:

1. *Protection forests* — it includes forests on hill slopes, watersheds or rivers, riverbanks, sea-shores and other localities vulnerable to erosion and degradation.
2. *Production forests* — which are essentially the commercial forests comprising valuable or potentially valuable timber-bearing stands occuring in favourable regions which are indispensable for the development of the country and for meeting the diverse requirements of the national economy.
3. *Social forests* — these would cover waste lands, panchayat, village, common lands on sides of roads, canal-banks, railway lines etc.

   The objectives of these forests will be :

   (a) agricultural timber and fuelwood,

(b) grazing and grass,
(c) recreation.

Weightage to Production Forests

The report has much to say about 'production forestry', as it is for industrial production, supplemented by adequate forest-based industries for which necessary institutional and technical changes should be made. These forests are expected to be project oriented and commercially feasible from the point of view of costs and returns. Among the several measures recommended, one is that, " . . . export of wood-based products should be stepped up between now and 1980, so that if any import of any class of wood-based product becomes necessary after 1980, it should be more than offset by exports".[37]

Further sub-division of production forests was recommended as :

(a) mixed quality forests,
(b) valuable forests, and
(c) inaccessible forests, making sure that these forests were not used for any other purpose but commercial.

The Commission also suggested that, " . . . identification of the areas where production forestry is possible should be done, particularly in heavy rainfall areas, and the priority should be on inaccessible forests followed by mixed quality and valuable forests".

It is clear that the highest priority is given to the production forests and lowest to the social forests. There is a suggestion that unclassed and protected forests should be constituted into reserved forests at the earliest, in order to extinguish nistar rights of the tribals, as far as possible.

In order to remove the hindrance of State boundaries in the transport of timber, the Commission comments that, " . . . the planned development of industries like pulp and paper would require carving out, beforehand, an industrial catchment which may transcend the borders of the States". It also recommended strengthening legislation on forestry for effective implementation of forest policy and enactment of a revised All India Forest Act. The most criticised draft Bill of 1980, which was subsequently dropped, was based mainly on the recommendations of the National

Commission on Agriculture.

But before that, two major policy decisions were taken which too were based on the recommendations of the said Commission. One was establishment of the Forest Development Corporations and second the promotion of social forestry activities.

FOREST DEVELOPMENT CORPORATION

The Necessity

The sectoral analysis of the nation's forest and forest-based-industries was carried out in 1971 by the Ford Foundation and the Planning Commission. The conclusion of the analysis was that India's industrial forestry programme must be geared up to increase production from the then level of 8.92 million cu.m. to 25.55 million cu.m. by 1980 and to 41.00 million cu.m. in 1990. The problems had also been studied by the National Commission on Agriculture which reported that forestry will not be contributing its due share to the national income unless the per hectare yield was increased.

The Commission was convinced that a new commercial approach to forestry development had to be built up. It noted that, "The average gross revenue from forests in India is Rs. 21.50 per hectare, whereas in Austria it is Rs. 336, Rs. 565 in West Germany and Rs. 494 in Switzerland".[38] Therefore, a change-over from conservation forestry to a more dynamic programme of production forestry was found necessary.

It was believed that the under utilisation of forest resources and its resultant meagre contribution to the national GNP (1.6%) and employment (0.2%) was largely due to the low rate of public investment in forestry and forest industries' development, absence of institutional finance and rigid organisational system within which forestry had to operate. Therefore, the Commission underlined that higher investments will automatically give job opportunities for a large number of skilled and unskilled personnel in the forest areas and generate further employment in the secondary and tertiary sectors of input production and services.

Objectives of the Forest Development Corporations

The concept of Forest Development Corporations was evolved with the specific objectives of —

(i) making use of industrial finance;
(ii) ploughing back substantial portion of profits for development;
(iii) giving flexibility and freedom to managers to introduce innovations and take quick and timeless decisions; and
(iv) to diversify production and to induct specialists in the organisations.

It is peculiar that there is no mention of simultaneous and integrated development of tribals.

In the State of Maharashtra, forest comprises 20.8 per cent of the total land expanse, covering 60,955 sq. km. A little over half of this area is unproductive or under-productive consisting either of land covered with sparse tree growth or grass lands or eroded barren lands devoid of vegetative cover. Thus the productive and potentially productive forests form a bare 10.00 per cent, from which the ever increasing demand for forest produce is to be met.

The Government of Maharashtra had set up a semi-autonomous Forest Development Board in March 1969, which was merged into the Forest Development Corporation of Maharashtra (hereinafter FDCM) in October 1974.

Objectives of FDCM

The basic objectives of the FDCM are :

(i) to increase the productivity of forests by replacing low value forests with high yielding forests through planting valuable tree species;
(ii) to strengthen rural base of economy by promoting greater employment opportunities to rural populace, and
(iii) to promote welfare of *adivasis* and other people below poverty line in rural areas.

Principal Activities

The Government of Maharashtra has set apart an area of 4660.472 sq. km of naturally grown forests to be covered in thirty years with valuable species by FDCM. FDCM's principal activities are:

(i) Clearance of existing stands of natural forests over approximately 12,000 hectares annually; and

(ii) Raising plantation of teak and other economically important species such as bamboo, *khair* etc. on the cleared lands.

It is interesting to note that, though strengthening the base of rural economy and welfare of tribals is one of the objectives of the corporation, it is not even mentioned in the activities to be undertaken.

However, clearfelling is totally stopped and presently FDCM has been given the job of wasteland development. (Further discussion on the Corporations' activities, is given in the Chapter 3).

SOCIAL FORESTRY

Under the double crisis of unsatisfied basic needs of tribals on the one hand and ecological instability on the other, the rebuilding of India's forests wealth has become one of the major issues in land-use policy, in recent years. Social forestry scheme is supposed to provide a new paradigm as a new strategy used to regenerate forest resources through the participation of the community in the protection and management of forests.

There are two inter-related concepts involved in social forestry. Firstly, social forestry means production of fuelwood, fodder and small timber for the use of rural communities. Secondly, social forestry is denoted by the land it covers — wastelands, panchayat lands, village woodlots, and lands on the sides of roads, canal banks and railway lines which may be brought under tree plantations. In other words, lands which are not under the direct control of Forest Departments, but are public and communal lands are covered by the social forestry.

Social forestry programmes have mainly three components : farm forestry, encouraging farmers to plant trees on their own farms by distributing free or subsidised seedlings; woodlots planted by the forest departments for the need of the community, especially along roadsides, canal banks and other such public lands; and community woodlots planted by the communities themselves on community lands, to be shared equitably by them.

Where social forestry differs most radically from past forestry programmes is in the recognition that the rebuilding of India's forest wealth cannot be undertaken without the participation of the local communities, "Community forestry cannot be imposed from above and carried out in the face of a hostile population. New forms of landuse impinge upon, and are influenced by the daily activities of everyone. When the local people are not active participants and supporters of a project, saplings have a way of disappearing overnight. With fodder usually as scarce as firewood, uncontrolled goats or cattle can quickly ruin a new plantation even when the disgruntled peasants are forced to face the alternative of a lengthy hike to collect fuel. Community involvement, then, is not just an ideologically appealing goal; it is a practical necessity if rural forest needs are to be met".[39]

(Detailed discussion and evaluation with its impact will be undertaken in the next chapter.)

The Indian Forest Bill, 1980

It was expected that the forest Bill drafted for the first time in free India would be different than the earlier efforts by giving due weightage to the dependence of tribal communities on forests. However, social activists working among the forest dwelling communities saw in the Draft Bill an " . . . increased threat to the life support system of these communities".[40]

The Bill tried to formulate certain definitions very widely. Thus, 'forest' included any land containing trees or shrubs, pasture lands and 'any land whatsoever', which the State Government may by notification declare to be forest for the purpose of this Act. Similarly 'forest produce' was defined as to include trees, leaves, flowers, fruits, all other parts or produce of trees not specifically mentioned, as well as all types of grass, creepers, orchids, moss and all parts or produce of such plants.[41]

The Bill classified forests into three categories: reserved forests, protected forests and village forests. Reserved forests imply 'prohibited areas' wherein the local forest dwellers and adiwasis are barred entry and the normal age-old practice of looking to the forest for the satisfaction of their basic needs were to attract a stiff penalty. The prohibited actions such as grazing of cattle, utilising bark for medicine and fibre for fastening houses, wood for implements and housing, leaves for plates, cups and storage, collection and barter of seeds, gum, honey, fishing and hunting small games — were labelled as 'offences' liable to a penalty or fine, imprisonment as well as payment of compensation for damage.

To the social workers, it was a cause of concern to note that "... while the Indian Penal Code prescribes the maximum possible penalty for offences, the Forest Bill lays down minimum penalty as well ... to a population that has no access to the Gazette and cannot even read or write".[42]

Under section 36, the Bill gave to the State Government the authority to regulate pasturing, hunting, fishing etc. in 'protected forests' as well. On village forests there was no concrete provision that spelt out how lands marked out for village forestry could be useful to the local population. It only said that the State Bureaucracy will be responsible for the manner in which the local population should live in harmony with the forest.

Chapter XII, Section 118 and 119 of the Bill gave judicial powers to forest officers to arrest and seize the property. It also gave forest officers *discretion* (emphasis added) in the collection of fines and calculations of compensation for damage.

The provisions in the Bill recognised personal claims over lands and forest produce, and ignored the fact that many of these rights especially those connected with forest produce have been enjoyed by the tribes as groups. Thus the Bill emphasised protection of forests *from the people* (emphasis added). It denied the real worth of forests in being a source of livelihood for millions of the tribal poor.

The Bill, without any change in basic attitude, became the Forest Conservation Act, 1980. Facing the nationwide uproar in 1982 against the Act, the Government decided to prepare a forest policy document. The new forest policy document was prepared by the end of 1983.

It was revised in 1987. But both the draft policies were criticised by the ecologists, eminent scholars and activists, as both the drafts incorporated nothing more than cosmetic changes to the 1952 policy.

National Forest Policy, 1988

The National Forest Policy resolution, 1988 is strikingly different from the previous draft statements. It reflects the pro-tribal attitude with clarity. Though it relates the rights and concessions of the tribals to the carrying capacity of forests, it also mentions the 'need for optimising this capacity by increased investment, silvi-cultural research and development of the area'. It assures that 'customary rights and concessions are fully protected', adding that 'the domestic requirements of fuelwood, fodder, minor forest produce and construction timber should be the first charge on forest produce'.

Other salient features of the policy resolution are :

(1) Accepting the highly depleted state of forests, it attributes this to the tendency to work upon forests as revenue earning resource.

(2) The basic objectives of the policy incorporates 'encouraging utilisation of forest produce and maximising substitution of wood'.

(3) It emphasises programme of afforestation with special emphasis on augmenting fuelwood production to meet the requirements of the rural people.

(4) One of the 'essentials' of the policy document is "minor forest produce provides sustenance to tribal population and to other communities residing in and around the forests. Such produce should be protected, improved and their production enhanced with due regard to generation of employment and income".

(5) Green belts in urban and industrial areas, timebound programme for fuel and fodder development, revenue from such programmes to be shared with local communities and creation of certain ownership rights over trees, are some of the remarkable suggestions.

(6) Section 4.6 deals with 'Tribal People and Forests'. It accepts the symbiotic relationship between tribals and forests and need to associate tribals in the protection, regeneration and development of forests.

(7) It is suggested that Forest based industries should grow their own raw material under the technical guidance of Forest Department. Also, these industries should not only provide employment to local people on priority but involve them fully in raising trees and raw material as well.

Although lot more could have been done to make it a right forest policy in all sense, it still stands as a good beginning.

In Maharashtra

The Government, after the formation of the State in 1961, adopted the Indian Forest Act, 1927, in toto. Very few changes have been made through amendments, since then. For example, the Maharashtra Felling of Trees (Regulation) Act, 1964 for the purpose of the preservation and protection of the soil against erosion. In 1964 itself an amendment in the Indian Forest Act was made regarding grant of licences to arrest the large scale illicit cutting of trees and conversion thereof into timber in the saw mills located within the forest limits. Subsequent amendments in this regard were made in 1965 and 1984.

Summary

It is clear from the above discussion that the forest-dwelling tribal communities had free access to forests and depended completely on forests before the advent of the British.

The British Government converted the natural resource into a supplier of raw-material for the forest based industries and a source of revenue for the government. Their forest-policy in the form of forest legislations resulted in increasing restrictions on the tribals and their use of the forest-produce.

The policy adopted by the Indian Government is in continuation of the Britishers' revenue-oriented approach. Forests continue to be 'raw material for industries' and 'income for government'. The people living in forests still continue to be treated as cheap labourers — living on the mercy of the forest-department officials. Every new programme and project in forests, adopted by the government provides an evidence of this. Even after severe criticism from scholars, social workers etc., the government seem to be bound by some 'other consideration', rather than the tribal welfare.

NOTES

1. McLellan David, *Marx Before Marxism*, The Macmillan Press Ltd., G. Britain, p. 96.
2. Sharma, B.D., 'Tribal Economy and Regional Development', Occasional Papers on Tribal Development, Ministry of Home Affairs, Government of India, New Delhi, 1978, p. 2.
3. Gadgil, M., 'Forestry with a Social Purpose', Towards a New Forest Policy, Indian Social Institute, Lodhi Road, New Delhi, 1983, p. 115.
4. Roy Burman, B.K., 'Issues in Environmental Management Centering Forest and Role of Tribal Communities', *South Asian Anthropologist*, 1985, p. 43
5. *Ibid.*, p. 44

6. Guha, R., 'Forestry in British and Post-British India — A Historical Analysis', *Economic and Political Weekly*, Vol. XVIII, No. 44, Oct. 29, 1983, p. 1883.
7. Gadgil, M., et al., 'Whither Environmental Activism', *Deccan Herald*, Feb. 19, 1984, Bangalore.
8. Guha, R., *op. cit.* p. 1884.
9. Prabhu, P., 'Social Forestry : An Adiwasi Viewpoint', Towards a New Forest Policy, *op. cit.* p. 134.
10. 'The Old Forest Policy', Circular No. 22-F, Oct. 19, 1894. Hundred Years of Indian Forestry, Forest Research Institute, Dehradun, 1961, p. 337.
11. 'The Government Forest Act, 1865', Hundred Years of Indian Forestry, *op. cit.*
12. *Ibid.*
13. 'Indian Forest Act, 1878', Chapter I, section-2, Government of India.
14. *Ibid.*, Chapter 2, para-3.
15. *Ibid.*, para-6(c).
16. Gadgil, M., 'Whither Environmental Activism', *op. cit.*
17. *Ibid.*
18. Guha, R. op. cit. p. 1884.
19. Kannan, K.P., 'Forestry Legislation in India : Its Evaluation in the Light of the Forest Bill, 1980', Towards, New Forest Policy, *op. cit.* p. 76.
20. Kulkarni, S.D., 'Towards a Social Forest Policy', *Economic and Political Weekly*, Vol. XVIII, No. 6, Feb. 5, 1983, p. 192.
21. 'The Old Forest Policy', *op. cit.*, para-5.
22. Kulkarni, S.D. *op. cit.*, p. 193.
23. Report of the Scheduled Areas and Scheduled Tribes Commission, 1960-61. Government of India, Manager of Publications, New Delhi, p. 21.
24. *Ibid.*, p. 21.
25. Kulkarni, S.D., *op. cit.*, p. 193.
26. 'National Forest Policy, 1952', Hundred Years of Indian Forestry, *op. cit.*, p. 329.
27. *Ibid.*, p. 329.
28. *Ibid.*, p. 331.
29. *Ibid.*, p. 333.
30. Gadgil, M., 'Whither Environmental Activism', *op. cit.*
31. Report on the National Commission on Agriculture 1976, Part IX 'Forestry', Government of India, Ministry of Agriculture and Irrigation, New Delhi, pp. 32-33.
32. *Ibid.*, pp. 32-33.
33. Report of Scheduled Areas and Scheduled Tribes Commission, *op. cit.*, p. 132
34. Report of the National Commission on Agriculture, 1976, *op. cit.*, Appendix 41.1.11.
35. 'Criminal Apathy of Maharashtra and Madhya Pradesh Governments — Contractor, Trader, Official clique denude border forest', *Nagpur Times*, 14th March, 1988, Nagpur.
36. Report of the National Commission on Agriculture, 1976, *op. cit.* Appendix 41.7.5.
37. *Ibid.*
38. Chowdhary, K., Managing Forests for Development', Centre for Science and Environment, Nehru Place, New Delhi p. 2.

39. Eckholm, Erik, 'Planting of the Future Forestry for Human Needs, Worldwatch Institute, Washington D.C. p. 1979.
40. D'Abreo, Desmond, 'People and Forests : The Forest Bill and a New Forest Policy', Indian Social Institute, New Delhi, 1982.
41. Draft 'Indian Forest Bill, 1980', Section 2(3), (6), Government of India, New Delhi.
42. 'Statism Not an Answer', *Economic and Political Weekly*, Vol. XVI, No. 36, Sept. 5, 1981, p. 1447.

3

IMPACT OF FOREST POLICY

It is clear from the discussion in Chapter 2 that the acts and policies adopted right from the British period were formulated for enhancing commercial use of forests by supporting industries and by ignoring the rights of the tribals on the forests. The Report of the Scheduled Areas and Scheduled Tribes Commission (1961), gives the gist of the British policy as follows :

> The sole object with which State forests are administered is the public benefit. In some cases, the public to be benefited are the whole body of tax-payers; in others, the people of the tract within which the forest is situated; but in almost all cases the constitution and preservation of a forest involve, in greater or lesser degree, the regulation of rights and the restriction of privileges of user in the forest area which may have previously been enjoyed by the inhabitants of its immediate neighbourhood.

Commercial Interests: Reign Supreme

Forest produce had only limited commercial importance till the British started laying an extensive network of railway lines using wooden sleepers. Between 1853 and 1910, 51,658 km of track was laid which meant increase in the revenue of the Forest Department from Rs. 0.05 millions to Rs. 239.90 millions for the same period. It is also worthy to note that between 1860 and 1910 though the kilometrage increased by 50 times, the increase in the net earnings of the railways increased by nearly 79 times (Table 3.1).

With this new form of commercialisation, forests began to be viewed primarily as a source of revenue and of industrial raw materials. Today, forests support several industries such as paper, plywood, polyfibre, matches, turpentine, drugs etc. But, at the same time, it is not adequately appreciated that forests are also the very basis of survival for more than 5 crores of tribals in India, even today.

TABLE 3.1: Railway Expansion in India (1853–1910)

Sl. No.	Year	Kilometrage	Outlay (Rs. million)	Net earnings (Rs. million)
1.	1853	32	3.8	0.05
2.	1860	1,349	266.6	3.00
3.	1870	7,678	900.0	30.30
4.	1880	14,745	1,285.7	63.90
5.	1890	26,395	2,136.7	103.60
6.	1900	39,834	3,296.1	164.50
7.	1910	51,658	4,390.5	239.90

Source: 'History of Indian Railways', Government of India, Delhi, 1964, p. 214.

Tribals as Intruders

The forest dwellers, majority of whom is formed by tribals, predominantly depend on forests for their survival, from times immemorial. A study of the efforts made by the British Government to meet the two primary requirements of land and forest, of tribals, reveals that as they proceeded from a policy of isolation and *status quo*, the tribals had to remain at the mercy of petty officials and nothing positive by way of rehabilitation and development was achieved. With increasing commercialisation, tribals became a hindrance in the eyes of the British Government which termed them as intruders.

Independent India's Priorities

Before formulating the Forest Policy in 1952, the Union Government took note of the following factors:

(i) The population of the country had increased by 40 per cent between 1894 and 1950 resulting in relentless pressure on forests and wastelands to secure more land for agriculture;
(ii) The importance of forests in mitigating the rigours of climatic conditions had come to be better understood;
(iii) The importance of forests in the economic field, such as the development of agriculture, industry, and communications was better understood; and
(iv) The two great world-wars brought to the fore hitherto unsuspected dependence of national defence on forests.[1]

The Indian Forest Policy, 1952, was based on the six paramount needs (already mentioned in Chapter 2), which gave maximum importance to environmental needs, revenue, production and exploitation of timber and other forest produce required for defence, communications and industry. But there was no mention of tribals' dependence on forests and only few provisions were made to meet their requirements.

In this chapter, the impact of the forest policy is studied in two sections. Section I deals with the impact on environment, and Section II deals with the direct and indirect effects of the policy adversely affecting the tribals.

SECTION I

IMPACT ON ENVIRONMENT

Top Priority to Ecological Balance

In 1952, India had only about 20 per cent of the total land area under forests. This was considered very inadequate and it was laid down that the average proportion of the forest area should be raised to 33.33 per cent. To provide protection against erosion, floods and denudation in the Himalayas, the Deccan plateau and other mountainous regions, it was stated that the forest area in those parts should not be allowed to fall below 60 per cent.

But in reality, " . . . there is probably no other area of India's environment that has been more viciously attacked and destroyed in the last century than the country's forests".[2]

Dwindling Forest Cover

Forest Department statistics has repeatedly pointed out that approximately 23 per cent of the country's total area i.e. 75 million hectares, is classified as forest land. But what the statement actually means is that this much area is under the control of the forest departments. There is no guarantee that it has tree cover. For example in Maharashtra, in 1980, Forest Department had 6.41 million hectares of area under its control but in reality only 3.04 million hectares was under tree cover. It is noteworthy that except in the States and Union Territories of Sikkim, Andaman and Nicobar and Arunachal Pradesh, the actual tree cover in all other States and

Union Territories has substantially reduced. For the States of Gujarat, Haryana, Himachal Pradesh, Jammu and Kashmir, Maharashtra, Punjab, Rajasthan and Dadra and Nagar Haveli, the reduction in actual forest area was more than 25 per cent from 1972-75 to 1980-82 (Table 3.2).

The satellite data released in mid-1984 by the National Remote Sensing Agency (NRSA) has revealed that during 1972-75, only 16.89 per cent of the land actually had forest, which went down to 14.10 per cent in 1980-82. It shows that India lost 1.3 million hectares of forest every year in approximately seven years period between 1972-75 and 1980-82 (Table 3.3). What is more important to note is that it is as much as nine times, the annual deforestation rate of 0.15 million hectares claimed by the forest department.

Significant Findings Regarding Area under Tree Cover

The NRSA study breaks down the forest cover into three categories:

(i) Closed Forests, (the forest area which has good density of tree cover), i.e., at least 30 per cent of land covered by canopy;
(ii) Open or Degraded Forests, i.e., the forest area with very low density of tree cover, including wasteland; and
(iii) Mangrove Forests, i.e., the forests on coastal regions.

The study, while defining the closed forests, considers the forests whose canopy covers 30 per cent of the ground cover as compared to 40 per cent of cover mentioned in international studies.

According to NRSA study, during the 1972-75 and 1980-82, closed forests dropped from 14.12 per cent to 10.96 per cent, while degraded forests increased from 2.67 per cent to 3.06 per cent and Mangrove forests also dropped from 0.099 per cent to 0.081 per cent of the total land area.

Thus, during 1972-75 — 1980-82, India lost 10.4 million hectare (22.40%) of closed forests and 63,000 hectares (21.21%) of mangroves. And 1.29 million hectares (14.70%) of closed forests got converted into degraded forests (Table 3.3). It should be noted that, actually the extent of degraded forests would be much higher and that of closed forests would be much lower if the international classification of forests had been followed by the NRSA study.

TABLE 3.2: State-wise Forest Areas by Satellite (1980)

(in million hectares)

Sl. No.	State/Union territories	Forest area by satellite data (1972-75)	Forest area by satellite data (1980-82)	Percentage change from 1972-75 to 1980-82	Area controlled by Forest Department (1980)
(1)	(2)	(3)	(4)	(5)	(6)
1.	Andhra Pradesh	4.90	4.04	−17.55	6.41
2.	Assam	2.11	1.98	−6.16	3.07
3.	Bihar	2.27	2.01	−11.45	2.92
4.	Gujarat	0.95	0.51	−46.31	1.95
5.	Haryana	0.08	0.04	−50.00	0.16
6.	Himachal Pradesh	1.51	0.91	−39.73	2.21
7.	Jammu & Kashmir	2.23	1.44	−35.42	2.19
8.	Karnataka	2.95	2.57	−12.88	3.79
9.	Kerala	0.86	0.74	−13.95	1.11
10.	Madhya Pradesh	10.86	9.02	−16.94	15.39
11.	Maharashtra	4.07	3.04	−25.32	6.41
12.	Manipur	1.51	1.38	−8.60	1.52
13.	Meghalaya	1.44	1.25	−13.19	0.86
14.	Nagaland	0.82	0.81	−1.21	0.29
15.	Orissa	4.84	3.94	−18.59	6.77
16.	Punjab	0.11	0.05	−54.54	0.24
17.	Rajasthan	1.13	0.60	−46.90	3.49
18.	Sikkim	0.18	0.29	+61.11	0.26
19.	Tamil Nadu	1.67	1.32	−20.95	2.18
20.	Tripura	0.63	0.51	−19.04	0.59
21.	Uttar Pradesh	2.59	2.10	−18.91	5.14
22.	West Bengal	0.83	0.65	−21.68	1.18
23.	Andaman & Nicobar	0.33	0.64	+93.93	0.71
24.	Arunachal Pradesh	5.14	5.21	+19.72	5.15
25.	Dadra & Nagar Haveli	0.02	0.01	−50.00	0.02
26.	Goa, Daman, Diu	0.12	0.11	−8.33	0.11
27.	Mizoram	1.39	1.20	−13.66	0.71

Source: National Remote Sensing Agency, Quoted in 'State of India's Environment', 1984, A Citizen's Report, Centre for Science and Environment, New Delhi.

Purpose-wise Loss of Forest Area

Despite the indication and guidelines given by the forest policy of 1952, the forest cover dwindled, instead of increasing to 33.33 per cent. If a purpose-wise forest area lost between 1951 and 1972 is considered, it indicates the loopholes in the implementation of the policy.

TABLE 3.3: Forest Cover in India

(million hectares)

Sl. No.	Class of forest	Years 1972-75	Years 1980-82	Percentage change from 1972-75 to 1980-82
1.	Closed forests	46.42 (14.12)	36.02 (10.96)	−22.40
2.	Open forests	8.77 (2.67)	10.06 (3.06)	+14.70
3.	Mangrove forests	0.33 (0.09)	0.26 (0.08)	−21.21
	Total forest cover	55.52 (16.89)	46.35 (14.10)	−23.72

Note: Figures in brackets denote percentage to the total land area of the country.
Source: National Remote Sensing Agency, Quoted in State of India's Environment, op. cit.

The study of 'Forest Resources of Tropical Asia' undertaken by Food and Agricultural Organisation (FAO) reveals that in India the highest, i.e. 71.51 per cent forest was destroyed for agricultural purposes and 11.79 per cent for River Valley projects. The percentage loss due to Roads and Communication was 1.61. For the establishment of Industries it was 3.67 per cent and 11.42 per cent loss was for miscellaneous purposes (Table 3.4). This was done against the norm of conservation, set by Government of India itself.

The consequences of this colossal deforestation are the increasing floods, soil erosion, heavy siltation of dams and changes in microclimate. In other words, " . . . a progressive depletion of the country's ecological bank, is driving the country incessantly towards bankruptcy; flood damages alone being Rs. 1,000 crores every year".[3]

Serious Ecological Imbalances

The extent of forest cover is a good indicator of the health of the land. The country can hope to achieve ecological security only by increasing the vegetal cover to tackle the main problems of land degradation, namely:

(a) According to an estimate made by the Ministry of Agriculture in March, 1980, as much as 175 million hectares, out of the country's total land area of 304 million hectares for which record exists, are subject to environmental problems. 150 million hectares (85.71%) of country's land is affected by serious water and wind erosion, 6 million hectares (3.42%) by water-logging and 3 million hectares (1.71%) by shifting cultivation. 4.50 million ha (2.57%) is affected by saline soil, 2.50 million ha (1.42%) by alkaline soil. 6.60 million ha (3.80%) land is classified as 'other cultivable waste land fit for reclamation, and 2.40 million ha (1.37%) as Diara Land (Table 3.5).

TABLE 3.4: Forest-Area Lost Between 1951 and 1972

(*thousand hectares*)

S.No.	Purpose	Area	Percentage loss
1.	River valley projects	401	11.79
2.	Agricultural purposes	2,433	71.51
3.	Roads and Communication	55	1.61
4.	Establishment of Industries	125	3.67
5.	Miscellaneous	388	11.42
	Total	3,402	100.00

Source: Forest Resources of Tropical Asia, FAO, 1981.

(b) In a study made in 1972, it was estimated that on an average, India was losing about 6,000 million tonnes of top soil per annum through water erosion and that these represented, in terms of major nutrients, an annual loss of Rs. 700 crores.[4]

(c) Again, according to the Report of the National Commission on Floods (1980), the losses on account of floods in 1976, 1977 and 1978 were Rs. 889 crores, Rs. 1,200 crores and Rs. 1091 crores, respectively. (It is noteworthy that according to an unpublished dissertation submitted for M.Phil degree in Nagpur University by Ms. V. Arvinda, the damages by the floods only in one year 1986 — and only by one river — Godawari, in one district — West Godawari, were Rs. 725.48

crores). According to the same report, the total area subject to periodic floods which was estimated at 20 million hectares in 1971, now stands at the level of 40 million hectares — an increase of 100 per cent in 10 years.

(d) The Himalayan eco-system has considerably deteriorated, resulting in floods in the Indo-Gangetic plains thereby causing heavy damage to property and crops resulting in untold human misery. Since, the run-off of rainwater from denuded areas is far greater than from well-wooded slopes, a great deal of water which could otherwise have been retained as subsoil and ground water, is today being lost as surface run-off, often causing further erosion and floods in the process.

TABLE 3.5: Land-Area with Environmental Problems

(*in million hectares*)

S.No.	Problems	Area	Percentage to total
1.	Serious water and wind erosion	150.00	85.71
2.	Shifting cultivation	3.00	1.71
3.	Water logging	6.00	3.42
4.	Saline soil	4.50	2.57
5.	Alkaline soil	2.50	1.42
6.	Diara Land	2.40	1.37
7.	Other cultivable waste land fit for reclamation	6.60	3.80
	Total	175.00	100.00

Source: Roy Burman, B.K., Issues in Environmental Management Centering Forest and Role of Tribal Communities, *South Asian Anthropologist*, 1985.

Commercial Interests Supersede Environmental Interests

The Indian Government has always given high priority to the environmental needs of the economy. But, now it has been established that in reality it is commercial interests which got benefitted.

Even after the clear mention in the Forest Policy of 1952, and conversion of all major forest cover into Reserved or Protected forests, the Government has not achieved the target figure of 33.33 per cent of total geographical area under tree-cover. Instead, it has further exploited whatever forest was left behind by the British. There are quite a few policy decisions which are contradictory to the

Government's policy of enhancing forest cover in the country. The purpose of establishing the Forest Development Corporations, unbelievable concessions to paper industries etc. are good examples. This revenue-oriented approach has always ruled supreme and the Indian Government, has, time and again, given-in to the pressure from the industrialists' lobby in the name of development, at the cost of the natural wealth and those who depend on it. (This revenue-oriented policy will be discussed in details under Section II of this chapter).

In the case of the big irrigation projects as well, the benefits ensuring from them are highlighted while the loss in terms of tree-cover and displacement of people therefrom is totally disregarded.

Tribals, the forest dwellers, are always accused of illicit felling, misuse of concessions, age-old harmful habits, and are held responsible for dwindling forest cover. But, one thing which is lost sight of is that tribals have always lived in forests and the problem of deforestation never occurred before the commercial use of timber became well-known. Curtailment of rights and privileges of tribals on the one hand and the disappearance of forests on the other, have really left the tribals in a sorrowful plight. The worst affected class due to the denudation is that of tribals, which is also the most neglected one.

In brief, though the Government of free India always gave top priority to conservation in the policy statements, in practice it could never achieve the goal set by itself. The aspect of environment was totally neglected which resulted into several ill-effects of the ecological imbalance. Many other policy decisions gave much more importance to the commercial use of forests which resulted in the loss of tree cover. Government also neglected the tribals, who live in and on forests, blaming them and holding them responsible for the deterioration of forest wealth.

SECTION II

ADVERSE IMPACT ON TRIBALS

That the tribals are a neglected class, for the policy formulators, is proved by the fact that no policy statements till the one in 1988, and the text of the Indian Forest Act, 1927 uses the word 'Tribal'. The six paramount needs of the country, on which the 1952 policy was based, do recognise the need for grazing and small wood of the people in general but not of tribals in particular who live in and around forests, thereby shrugging off the responsibility of development of tribals in the backward areas.

The literature produced by Forest Departments highlights their efforts regarding scientific exploitation with minimum wastage, successful plantation of commercially valuable species, ever increasing revenue, increasing number of the forest-based industries, etc. But what is lost sight of is that in this endeavour tribal interests are adversely affected. This attitude of the policy formulators has resulted in largely reduced availability of forest produce for tribals, adversely affecting their standard of living, employment opportunities and displacement of tribals. Forest Departments, whose labour-force has a tribal majority, find tribals a liability because of their rights and privileges now converted into concessions.

In the following pages indirect effects of the policy are discussed first and the direct effects later on.

A. **Indirect Effects of the Policy at National Level**

(i) *Forest A Major Source of Revenue*

Though timber and other forest produce have been used since ancient times, forest produce gained considerable commercial importance only after Britisher's revenue-oriented outlook, which reflected through their policy. Forests became an important source of revenue for them and same is true even today.

Revenue of the forest department between the period 1869 to 1925 increased from Rs. 5.6 million to Rs. 56.7 million (i.e. ten time), while the surplus of the department increased from Rs. 1.7 million to Rs. 21.3 million (i.e. twelve times) over the same period (Table 3.6).

During seven years just before Independence, the revenue increased from Rs. 29.4 million to Rs. 124.4 million i.e. four times. The surplus increased from Rs. 7.2 million to Rs. 48.9 million i.e. six times. The out-turn of timber and fuel also nearly doubled in this period from 270 million cu. ft. to 439 million cu. ft. (Table 3.7). This was the period of Second World War, when the British defence depended on Indian teak to a large extent. Thenceforth, demand for timber and industries based on forest produce was ever on the increase.

TABLE 3.6: Revenue and Surplus of Forest Department. 1869-1925

(*Rs. million*)

Period	Revenue	Surplus	Percentage of Col. 3 to 2
1870-74	5.6	1.7	30
1874-79	6.7	2.1	31
1879-84	8.8	3.2	36
1884-89	11.7	4.2	36
1889-94	15.9	7.3	46
1894-99	17.7	7.9	45
1899-04	19.7	8.4	43
1904-09	25.7	11.6	45
1909-14	29.6	13.2	45
1914-19	37.1	16.0	43
1919-24	55.2	18.5	34
1924-25	56.7	21.3	38

Source: Compiled from 'Indian Forest Statistics', 1870-74 to 1924-25, Government of India, Delhi (1949).

TABLE 3.7: India's Forests During Second World War

(*Rs. in million*)

Year	Outturn of timber and fuel (m.cu.ft.)	Revenue	Surplus
1937-38	270	—	—
1938-39	299	29.4	7.2
1939-40	294	32.0	7.5
1940-41	386	37.1	13.3
1941-42	310	46.2	19.4
1942-43	336	65.0	26.7
1943-44	374	101.5	44.4
1944-45	439	124.4	48.9

Source: Compiled from 'Indian Forest Statistics', 1939-40 to 1944-45, *op. cit.*

Freedom to India did not bring about any major change in the outlook in the commercial use and revenue-oriented policy towards forests. Statewise contribution of forest revenue for the year 1980-81 clearly shows that all those states which have substantial forest areas also have substantial contribution by forests to the total revenue, for example, Himachal Pradesh (8.37%), Jammu and Kashmir (8.31%), Karnataka (5.04%), Kerala (5.37%), Madhya Pradesh (12.50%), Orissa (4.69%), Maharashtra (2.71%), Tripura (13.15%) and Andhra Pradesh (2.38%) (Table 3.8).

TABLE 3.8: Statewise Contribution of Forest Revenue in 1980-81

(Rs. '00 million)

Sl. No.	Name of the State	Revenue from all sources	Revenue from forests (Provisional)	Percentage contribution of forest to revenue from all sources
1.	Andhra Pradesh	109.283	2.609	2.38
2.	Assam	30.373	1.125	3.70
3.	Bihar	87.465	1.471	1.68
4.	Gujarat	94.750	1.128	1.19
5.	Haryana	43.332	0.159	0.36
6.	Himachal Pradesh	79.288	1.616	8.37
7.	Jammu & Kashmir	22.573	1.878	8.31
8.	Karnataka	91.244	4.600	5.04
9.	Kerala	59.129	3.178	5.37
10.	Madhya Pradesh	111.488	13.944	12.50
11.	Maharashtra	192.197	5.220	2.71
12.	Manipur	8.456	0.024	0.28
13.	Meghalaya	6.636	0.088	1.32
14.	Nagaland	9.543	0.050	0.52
15.	Orissa	53.690	2.521	4.69
16.	Punjab	54.472	0.404	0.74
17.	Rajasthan	71.659	0.651	0.90
18.	Sikkim	3.501	0.032	0.91
19.	Tamil Nadu	98.266	1.125	1.14
20.	Tripura	8.555	1.125	13.15
21.	Uttar Pradesh	162.234	2.030	1.25
22.	West Bengal	112.373	1.423	1.26
	Total	1,450.518	46.401	3.19

Source: The State of India's Environment, 1982, A Citizen's Report, Centre for Science and Environment, New Delhi.

For all these States the surplus of forest department is also substantial. Except Gujarat, Haryana, Manipur, Nagaland, Punjab, Rajasthan, Sikkim, Uttar Pradesh, Andaman and Nicobar, Goa and Mizoram, all other States and Union Territories get more revenue than expenditure from their forest departments. Madhya Pradesh and Maharashtra get surplus as high as Rs. 495.095 million and Rs. 252.328 million respectively (Table 3.9).

For the State of Maharashtra the surplus in the forest department is continuously increasing. Between 1970-71 and 1980-81 alone, the surplus increased from Rs. 3.37 crores to Rs. 22.95 crores (i.e. by 681%). It further increased to Rs. 50.27 crores in 1987-88. Out of the surplus for 1987-88 major contribution came from Amravati circle (Rs. 17.78 crores) and Chandrapur circles (Rs. 27.36 crores), both North and South Chandrapur circles put together (Table 3.10).

TABLE 3.9: Financial Working of Forest Department, 1980-81 Provisional (*million Rs.*)

Sl. No.	Name of the State	Revenue	Expenditure	Surplus
1.	Andhra Pradesh	260.914	148.249	+112.665
2.	Assam	112.455	63.530	+48.925
3.	Bihar	147.113	72.648	+74.465
4.	Gujarat	112.801	129.755	−16.951
5.	Haryana	15.897	30.933	−15.036
6.	Himachal Pradesh	161.585	130.645	+30.940
7.	Jammu & Kashmir	187.807	133.633	+54.174
8.	Karnataka	460.000	256.800	+203.200
9.	Kerala	317.836	76.788	+241.048
10.	Madhya Pradesh	1,394.412	889.317	+495.095
11.	Maharashtra	521.953	269.625	+252.328
12.	Manipur	2.393	10.560	−8.177
13.	Meghalaya	8.756	0.211	+8.545
14.	Nagaland	5.012	25.455	−20.443
15.	Orissa	252.067	85.613	+166.454
16.	Punjab	40.440	52.916	−12.476
17.	Rajasthan	65.134	150.251	−85.117
18.	Sikkim	3.156	10.659	−7.503
19.	Tamil Nadu	112.542	91.763	+20.779
20.	Tripura	112.469	92.471	+19.998
21.	Uttár Pradesh	203.000	233.318	−30.318
22.	West Bengal	142.338	131.671	+10.667
23.	Andaman & Nicobar	38.025	42.285	−4.260
24.	Arunachal Pradesh	38.181	23.330	+14.851
25.	Dadra & Nagar Haveli	2.259	1.733	+0.526
26.	Delhi	0.053	0.015	+0.038
27.	Goa	6.688	10.858	−4.170
28.	Mizoram	0.284	3.246	−2.962
	Total	4,725.560	3178.278	+1547.282

Source: The State of India's Environment, 1982, *op. cit.*

The composition of the revenue of Maharashtra Forest Department shows that the highest contribution has always been made by timber sale (74.21 per cent in 1960-61 and 64.47 per cent in 1980-81 and 69.23 per cent in 1988-89) which is followed by sales of

Bidi leaves (26.80 per cent in 1970-71 and 13.33 per cent in 1980-81 and 13.55 per cent in 1988-89). On the other hand, the contribution from other MFP (minor forest produce) has reduced substantially (i.e. from 7.73 per cent in 1960-61 to 1.09 per cent in 1980-81 and further to 0.73 per cent in 1988-89). The contribution from grass and grazing also has gone down from 3.13 per cent to 1.29 per cent and 0.91 per cent for the same period (Table 3.11). In case of firewood contribution has only marginally increased from 10 per cent to 11.39 per cent which has declined to 7.27 per cent for the year 1988-89. In case of Bamboo there is a gradual increase in its contribution. In 1961-62 it fetched 2.58 per cent of the total revenue which increased to 4.50 per cent by 1988-89. This strengthens the allegation that the forest department in commercially-oriented.

TABLE 3.10: Forest Revenue and Expenditure in Maharashtra

(*Rs. in crores*)

Sl. No.	Name of the circle/	Year	Revenue	Expenditure	Surplus
(1)	(2)	(3)	(4)	(5)	(6)
1.	Thane	1970-71	1.99	1.19	0.80
		1975-76	3.33	1.72	1.61
		1980-81	2.90	2.55	0.35
		1987-88	11.15	7.72	3.45
2.	Nasik	1970-71	0.93	1.06	−0.13
		1975-76	2.58	1.59	0.99
		1980-81	2.38	2.92	−0.54
		1987-88	4.04	4.35	−0.31
3.	Pune	1970-71	0.23	0.64	−0.41
		1975-76	0.41	1.07	−0.66
		1980-81	0.17	2.03	−1.86
		1987-88	0.22	3.47	−3.25
4.	Aurangabad	1970-71	0.24	0.65	−0.41
		1975-76	0.39	1.08	−0.69
		1980-81	0.62	1.18	−0.56
		1987-88	1.33	3.75	−2.42
5.	Amravati	1970-71	3.42	1.65	1.77
		1975-76	5.53	2.18	3.05
		1980-81	12.83	3.83	9.00
		1987-88	24.11	6.33	17.78
6.	Nagpur	1970-71	2.39	1.45	0.94
		1975-76	3.84	3.45	0.39
		1980-81	6.42	3.89	2.53
		1987-88	20.55	17.64	2.91
7.	Chandrapur	1970-71	4.70	2.63	2.07
		1975-76	7.10	3.24	3.86
		1980-81	15.16	7.00	8.16

(*Contd.*)

TABLE 3.10 (*Contd.*)

(1)	(2)	(3)	(4)	(5)	(6)
8.	Research & Education	1970-71	—	—	—
		1975-76	0.03	0.40	–0.37
		1980-81	0.12	0.59	–0.47
		1987-88	—	—	—
9.	Other Units	1970-71	0.02	1.49	–1.47
		1975-76	—	0.50	–0.50
		1980-81	0.02	0.90	–0.88
		1987-88	5.16	4.53	—
10.	Forest Board and FDCM	1970-71	1.24	1.03	0.21
		1975-76	—	—	—
		1980-81	15.39	8.17	7.22
11.	Dhule	1987-88	2.36	4.84	–2.48
12.	Kolhapur	1987-88	1.25	4.46	–3.21
13.	Yavatmal	1987-88	14.34	5.50	8.84
14.	N. Chandrapur	1987-88	11.57	6.57	4.00
15.	S. Chandrapur	1987-88	33.35	9.99	23.36
	Total	1970-71	15.16	11.79	3.37
		1975-76	25.82	15.67	10.15
		1980-81	57.29	34.34	22.95
		1987-88	129.46	79.19	50.27

Note: Dhule, Kolhapur, Yavatmal, N. Chandrapur and S. Chandrapur circles were carved out after 1980-81.

Source: Compiled from 'Annual Administration Report' Years — 1970-71, 1975-76, 1980-81, 1987-88, Forest Department, Government of Maharashtra.

Demand for timber and other forest produce, which is backed by good price, comes from the industrial sector. To make this demand perpetual, produce is given away at highly subsidised rates and to maintain continuous supply, the plantation of commercially useful species alone is taken up. Plantation of cheap small timber and grass, needed by tribals and other poor, is neglected. Tribals are not allowed to enter those plantations either for grazing their cattle, or for collecting firewood as that could harm the plantations. Hence, these forests with plantations are a permanent loss to tribals.

(ii) *Plantation of Commercial Species*

The revenue considerations of the government are also clear if one goes through the details of the area of established plantation. Though the policy of the government on paper gives more importance to environmental and fuelwood needs, the figures show a different trend.

Industrial plantation, which means direct revenue to the government, are preferred over any other plantation. Between 1951-80, industrial plantation in India occupied 69.4 per cent of the total area of established plantations. For all the thirty years, industrial plantation has had an upper age over fuelwood and environmental plantation. The area under 'much talked about' environmental plantation has dwindled from 25.19 per cent to 13.16 per cent . The area under plantation for fuel-wood increased from 1.16 per cent to 20.78 per cent, with the increasing demand for fuelwood (Table 3.12).

TABLE 3.11: Composition of Forest Revenue in Maharashtra

(*Rs. in lakhs*)

Sl. No.	Items	1960-61	1970-71	1975-76	1980-81	1988-89
1.	Timber	403 (74.22)	848 (55.79)	1480 (57.28)	2,704 (64.47)	7,774 (69.25)
2.	Firewood	54 (9.94)	118 (7.76)	205 (7.93)	478 (11.39)	817 (7.27)
3.	Grass and grazing	17 (3.13)	14 (0.92)	29 (1.12)	54 (1.29)	103 (0.91)
4.	Bamboo	14 (2.58)	36 (2.37)	98 (3.79)	127 (3.03)	506 (4.50)
5.	Bidi leaves	–	407 (26.78)	395 (15.29)	559 (13.33)	1,522 (13.55)
6.	Gum	–	–	–	20 (0.48)	–
7.	Hirda	–	–	–	7 (0.17)	–
8.	Other MFP	42 (7.74)	30 (1.97)	81 (3.14)	19 (0.45)	82 (0.73)
9.	Other receipts (mostly of timber & firewood)	13 (2.39)	67 (4.41)	296 (11.45)	226 (5.39)	424 (3.77)
	Total	543 (100.00)	1,520 (100.00)	2,584 (100.00)	4,194 (100.00)	11,228 (100.00)

Note: Figure in brackets denote percentage to total.

Source: Compiled from the 'Annual Administration Report', 1960-61, 1970-71, 1975-76 and 1980-81, 1990, *op. cit.*

As already discussed in the second chapter, social forestry means plantations basically to meet fuel and fodder requirements of the people. On the other, hand, production forestry is to supply raw material to the industries. However, the achievement of both, social forestry and production forestry prove the revenue considerations of the Indian Government. Right from 1952-56 (the First Five Year Plan) to 1969-74 (the Fifth five Year Plan) more than 70 per cent of plantation was done under 'production forestry' and 70 to 84 per cent of total funds were allocated towards it. Social forestry received only 12 to 30 per cent of the total funds and only 29 to 48 per cent of the total area was brought under plantation for this purpose (Table 3.13).

TABLE 3.12: Area of Established Plantation, 1951-80

(*'000 hectares*)

Sl. No.	Year/Type	Industrial	Fuelwood	Environmental	Total
1.	1951-60	190	3	65	258
		(73.65)	(1.16)	(25.19)	(100.00)
2.	1961-65	321	40	152	513
		(62.57)	(7.80)	(29.63)	(100.00)
3.	1966-70	534	68	142	744
		(71.77)	(9.14)	(19.09)	(100.00)
4.	1971-75	545	84	119	748
		(72.86)	(11.23)	(15.91)	(100.00)
5.	1976-80	607	191	121	919
		(66.05)	(20.78)	(13.17)	(100.00)
	Total	2,197	386	599	3,182
		(69.04)	(12.13)	(18.83)	(100.00)

Note: Figures in bracket indicate percentages.
Source: Forest Resources of Tropical Asia, FAO, 1981.

It was only after 1974, that the importance of plantation for fuel and fodder was underlined and in the Sixth Five Year Plan 70.94 per cent of the total plantation area was given under social forestry and 20.06 per cent was retained for production forestry.

Eucalyptus Frenzy

Eucalyptus is another good example of Forest Department's commercial attitude. Nearly half a million hectares have been brought under the exotic species as it means immediate returns, being a good raw material for paper industry. Out of 500 known eucalyptus varieties, 170 have been tried in India and five of them are planted on a large scale. The first large plantation work was taken up with *eucalyptus globulus* in 1956 in the Nilgiri Hills. Eucalyptus hybrid or *eucalyptus tereticornis*, popularly known as Mysore gum, has been most widely planted variety since the 1960s. The critics described it as an *ecological monster* for several reasons. Some of the reasons are discussed below:

TABLE 3.13: Achievements of Indian Forestry Since 1951

Sl. No.	Plan/Period	Social Forestry		Production Forestry		Total Plantation	
		Physical	Financial	Physical	Financial	Physical	Financial
1	2	3	4	5	6	7	8
1.	I Plan	15,130	1.642	36,957	11.191	52,087	12.833
	(1951-56)	(29.04)	(12.79)	(70.95)	(87.20)	(100.00)	(100.00)
2.	II Plan	147,222	19.933	163,948	48.688	311,170	68.621
	(1956-61)	(47.31)	(29.04)	(52.68)	(70.95)	(100.00)	(100.00)
3.	III Plan	260,315	54.305	323,054	156.964	583,369	211.269
	(1961-66)	(44.62)	(25.70)	(55.37)	(74.30)	(100.00)	(100.00)
4.	Annual Plan	127,806	42.890	325,171	187.346	452,977	230.236
	(1966-69)	(28.21)	(18.62)	(71.79)	(81.38)	(100.00)	(100.00)
5.	IV Plan	567,335	525.326	653,861	547.480	1221,196	1072.806
	(1969-74)	(46.45)	(48.96)	(53.55)	(51.04)	(100.00)	(100.00)
6.	V Plan	190,280	70.670	523,990	372.730	714,270	443.400
	(1974-79)	(26.63)	(15.93)	(73.37)	(84.06)	(100.00)	(100.00)
7.	1979-80	121,877	219.541	101,301	135.459	223,178	355.00
		(54.60)	(61.84)	(45.40)	(38.16)	(100.00)	(100.00)
8.	1980-81	152,811	337.242	95,325	161.259	248,136	498.501
		(61.58)	(67.65)	(38.42)	(32.35)	(100.00)	(100.00)
9.	Total	1582,776	1271.549	2223,607	1621.117	3806,383	2892.666
		(41.58)	(43.95)	(58.42)	(56.05)	(100.00)	(100.00)
	Targets	289,424	652.061	126,296	197.144	415,720	849.205
	(1981-82)	(69.61)	(76.78)	(30.39)	(23.22)	(100.00)	(100.00)
10.	VIth Plan	1523,870	3518.835	624,042	1003.124	2147,9122	4521.959
	(1980-85)	(70.94)	(77.82)	(29.06)	(22.18)	(100.00)	(100.00)

Notes: 1. Figures in brackets denote percentages.
2. Figures for physical achievement are in hectares.
3. Figures for financial achievements are in millions of Rupees.
4. Figures for 1981-82 and 1980-85 are provisional.

Source: Report of the Fuelwood Study, Planning Commission, (1982) Government of India, New Delhi.

(1) *Effect On Water-table*

A full grown eucalyptus tree consumes eighty gallons of water a day. Therefore it is not good for dry soils where nearby rivers and ponds go dry. Upper layer of soil which conserves water, nitrogen salts etc. gets eroded due to direct exposure to sun and wind. It intercepts less rainfall and allows soil eradication. A study carried out by the 'Soil and Water Conservation Research Institute' in Ooty (Tamil Nadu), found that interception in eucalyptus plantation was only 2.9 per cent, which is very low in comparison with other species. This allows easy run off of top soil. Eucalyptus hybrid is generally found in eastern coast of Australia which receives 700 to 1200 mm of rainfall. In low rainfall regions its roots spread nearly 13 metres preventing any undergrowth.

(2) *Soil Strength*

The soil-strength depends on what eucalyptus replaces and also on the planting density. If the density is above 10,000 trees per hectare, nitrogen and phosphorus deficiency may be the result. It, being a commercially paying species, is harvested in short rotation, thereby harming moisture and nutrients.

(3) *Undergrowth*

Many botanists point out that since lateral roots of eucalyptus hybrid and their ramifications occupy a portion of the top layer of soil, some species cannot compete with the root system and remain stunted if grown near old eucalyptus trees.

Now the question arises that why should any government encourage plantation of such species which has so many ecological drawbacks? Why is there a sudden forest-department frenzy for eucalyptus? Because, ". . . it is matchboxes, rayon, paper, furniture and medicine".[5] Profitability seems to be the only answer.

(iii) *Undue Concessions to Forest-based Industries*

With the help of undue concessions from the Government, forest-based industries increased in number and grew in dimension. Today, number of major and minor industries such as paper, plywood, drugs, matchbox, turpentine, tanins, furniture, construction, *bidi* manufacture etc. depend on forest produce. The production of the raw material for these industries means revenue to the Government. But this expansion has one major side-effect which is overlooked. Because of this liberal policy towards forest-based industries, tribals are being deprived of their small share of the forest-produce. Use of bamboo is one example taken here.

Bamboo : Paper Industry Versus Tribals

Paper is an important commodity and per capita paper consumption is often regarded as an index of literacy. After Independence, Government wanted paper industry to grow fast and hence gave the industry many concessions on raw material prices to encourage new establishments. State governments in their bid to attract industry, have often given away bamboo forests on lease at throwaway prices. The Orient Paper Mills at Amlai, Andhra Pradesh, was initially given the bamboo at 37 paise per tonne by the Madhya Pradesh Government.[6] As compiled in the 'State of India's Environment, Citizen's Report' (1984), the Seshasayee Paper Mills in Tamil Nadu was paying Rs. 6.89 per tonne until 1970, Rs 11.00 from 1970 to 75 and Rs. 22.00 plus 5.00 per cent for administrative expenses since 1975. Paper mills in Karnataka were paying Rs 15.00 per tonne for bamboo and as reported by some of the forest officials the paper mills in Dhule and Chandrapur in Maharashtra were paying between Rs. 7.50 to Rs. 10.00 per tonne.

Impact of this 'liberal' policy is felt now, when the capacity utilisation by paper mills is steadily decreasing from 98.90 per cent in 1970 to 68.00 per cent in 1981, and further to 58.1 per cent in 1983. (Table 3.14). The Mills produced 1.11 million tonnes in 1983, valued at Rs. 1,200 crores and used about 3.09 million tonnes of forest raw materials.

TABLE 3.14: Paper and Paper Boards Industry (1983)

Sl. No.	Item	
1.	Number of paper mills	175
2.	Annual capacity (m tons)	1.91
3.	Production (m tons)	1.11
4.	Capacity utilisation (%)	58.1
5.	Forest raw-material requirement (m tons)	3.09
6.	Per capita consumption (kg.)	1.65
7.	Estimated value of actual production	1,200

Source: Indian Paper Market Association, State of India's Environment, A Citizen's Report (1984) *op. cit.*

The supply of bamboo, the main raw material for paper, is shrinking. This has forced the producers to go in for unconventional raw materials like agricultural residues (rice and wheat straws etc.), cotton rags, hemp, jute cuttings and grass, or mix hard-wood with bamboo. In such crisis, both private and public sector paper mills are heading towards the last forest frontiers such as Bastar, the north-east and the Andamans.

The most important aspect of these concessions and subsidies is the treatment that the tribals got and the price which they had to pay for these materials. While the paper mills had to pay anything between 37 paise to Rs 22.00 for a tonne of bamboo, tribal craftsmen could purchase it in the open market only at Rs. 1,200 per tonne. The Buruds, the traditional basket weavers, found it impossible to buy bamboo at the market rate and could not sell their costly product at enhanced prices. Tribals, who were getting their fixed quota of bamboo as their 'nistar rights', ceased to get it because of non-availability. In Bhandara district of Maharashtra, the tribals had to protest by organising a *rasta roko* when they were denied their share of bamboo due to non-availability while truck-loads of bamboo were going to Ballarshah Paper Mill.[7]

The depletion of bamboo stocks in Tamil Nadu reduced the earnings of basket-weavers from Rs. 8.00 per day in the early 1970s to less than Rs. 6.00 per day now, in 1980's.[8] Furthermore, presently these basket-weavers are always under the threat of sudden periods of unemployment of several weeks due to non-availability of bamboo — a situation they had never known till a decade ago.

Bamboo is not just a raw material for basket weavers. It is a food item (bamboo shoots), an important hut building material and also a fuel. Some tribals even store water or grains in big bamboos and use it as utensils. Such a useful article for tribals, bamboo, is becoming all the more scarce because even now some state forest departments lease out bamboo forests to paper mills who leave behind nothing. While bamboo has disappeared from natural forest, the bamboo plantations in man-made forests do not recognise the rights of the tribals.

(iv) *Forest Development Corporations — Priority to Commercial Species*

The aims and objectives of setting Forest Development Corporations were clearly commercial as already pointed out in the second chapter. These corporations are to utilise the Forest resources fully to add to the GNP and to make use of industrial finance to replace the low value forests with high yielding forests. For this purpose, natural forests are cleared and teak, bamboo and other economically important species are planted. The Corporations also claim to be working for, ". . . strengthening of rural base and welfare of adivasis."[9]

Forest Development Corporations were set-up in many States and Maharashtra is one of them. All these corporations were given millions of hectares of natural forest which was clearfelled and commercially valuable species, such as teak and bamboo, were planted. All this was done for 'national development' by providing and assuring consistent supply of raw material to the forest-based industries. Lot of research is still being done to increase the yield, improve quality, and different combinations etc. are being experimented. But not even once did the government consider how this project would adversely affect the Tribals.

'Wages paid in return of labour used' was termed as tribal welfare and certain schemes under World Food Programme etc. were taken up. Under this scheme food articles, not usually used by tribals, were distributed.[10] Now the situation has worsened as all labour-intensive work of felling natural forests and plantation of commercially valuable species is almost over in many places in Maharashtra. Now, tribals not only find themselves out of employment but also have lost that particular forest area for ever. They have also ceased to get the 'food-packets' for which they are habitual now.

Also, how plantation of teak, and other 'commercially valuable species', which are of no direct use to tribals, would strengthen rural base, remains a point to ponder. Presently FDCM is given the work of wasteland development.

(v) *Social-Forestry*

As already mentioned in the second chapter, the major objectives of the development of social forests are to meet the demand of rural poor for fuel and fodder and to provide gainful employment to forest-dwellers and particularly to the *adivasis*. Social forestry is not merely growing of trees, but aims at growing trees with particular social aspect. This aspect has been well-emphasised by an eminent scholar Dr. K. Chowdhary in her paper 'Social Forestry for Integrated Development'. She rightly states that the objectives of social forestry are not physical but human. They are —

(1) to fulfil fodder, fuelwood, small timber demands.
(2) to provide employment opportunities and to increase family incomes considerably.
(3) to tap dormant energies and forestry skills of the village dwellers.
(4) to popularise economic tree-farming along with crop farming, and
(5) to integrate economic gains in distribution of benefits to the left-behind persons.[11]

"These aspects are often ignored by forest departments which think social forestry as growing of trees on lands outside their jurisdiction."[12]

Maharashtra Social Forestry project with a separate department was launched in 1982 with the total cost of $ 60 million (Rs. 60 crores approximately). The project received the loan and grants from the US AID.

The project was completed in September 1990 but it is too early to evaluate the success of the project. The aims and the other salient features of the project can be listed as follows —

(1) The goals of the project were given as —

  (i) an increased supply of firewood, fodder, fruit and small timber in rural areas;
  (ii) a reduced rate of deforestation; and
  (iii) an increase in rural employment.

(2) Land to be covered under this project is – the wasteland, land which is usually severely overgrazed and unproductive, government owned and community owned lands and also

planting of rail, canal and road-sides lands as well as marginal farmlands.

(3) The project is to cover approximately 4,300 villages in all the 27 rural districts of Maharashtra and 81,000 hectares will be covered in the project which will be managed by the communities.

(4) 'The area planted during the project will supply an estimated 43,000 villages and their 6.5 million inhabitants with fuel and fodder.'

(5) Grass production was to increase in the very first year due to the increased vigour of vegetation which was to be released from grazing pressure.

(6) Increased supply of fuelwood was assured.

(7) 'It will generate 30 million workdays of employment through activities related to ground preparation, planting, weeding etc'.

(8) 'Out of the total area planted, 90.00 per cent of the area would be covered by fuel and fodder species'.[13]

Since September 1990 the Social Forestry Department receives funds from the State Government and continues to function without any change in its objectives and programmes. However, the social forestry project was criticised for its industrial orientation, introduced under the populist guise of solving the rural energy crisis. It was quickly and rightly dubbed as a 'con game' by scholars who pointed to its long-term socio-ecological impact on rural population by growing species like eucalyptus and teak, and also on farmers who replaced their food crop by these species. "A close scrutiny of the scheme reveals that,

(a) they have emphasised plantation on private lands as against the stated objectives of plantations on village common-lands and degraded government-lands;

(b) they have favoured exotics like eucalyptus which are commercially valuable and which hardly fulfill 'basic' needs of fuel, fodder and fibre."[14]

It is these aspects that have led independent scholars to conclude that far from being a programme of the rural poor, social-forestry is essentially a response to the failure of production forestry.

According to another evaluation of the programme, social-forestry has succeeded with progressive farmers who could make commercial use of trees planted on their private lands. The plantation on roadsides, canal-banks and reforestation of degraded forest lands too have succeeded to some extent. But, "programmes focussed on the rural poor have not been able to generate their participation or enthusiasm although the village wood lot scheme stands to benefit the village community individually as well as collectively".[15] To the extent these afforestation plans have borne fruits, the success has been due to commercialisation of short duration plants and not by any other thing.

Another distortion that has crept into the scheme it its impact on employment. Originally conceived as an employment generation programme, it is not fulfilling this end. An evaluation study of the World Bank aided social forestry programme in Kolar district in Karnataka shows that while afforestation created 100 mandays of employment per hectare, loss of employment due to agricultural land being given away to farm forestry was 250 mandays per hectare. Thus, while the total programme was estimated to create 40 million mandays of employment over a period of five years, the loss of employment on the completion of the programme was reported to be as much as 137 million mandays in the same period.[16]

According to reports of an evaluation committee which were submitted to Maharashtra Government in 1988, the target of plantation over 81,000 hectares have completely failed. Out of several reasons given, the important ones are:

(1) plantations done on grazing lands which were not protected from the grazing cattle,
(2) plantations claimed only on paper, but nothing done in reality,
(3) mismanagement and large-scale corruption,
(4) reluctance of the forest officials to associate local population with the afforestation project,
(5) forest-guards were not appointed,
(6) no facility of watering the plantations was provided, and
(7) indifferent and irresponsible attitude of forest officials.[17]

It is clear from the foregoing discussion that social forçstry is not really meant for meeting the fuel, food and fodder requirements of the rural population, and also that the government policy is only serving commercial interests in disguise. It is true that failures of social forestry plantations cannot be totally attributed to the policy formulators but the elements of inefficiency and mismanagement along with corruption, do play an important role.

(vi) *Wasteland*

The increase in the extent of wasteland since Independence can be taken as a good indicator of the failure of the government policy.

India, today faces a peculiar situation of having over 129.58 million hectares (1.29 million sq.km) of its landmass classified as wasteland. It is approximately 39.41 per cent of the total geographical area of the country. The wasteland in the country comprises 936.91 lakh hectares of non-forest (72.30%) and 358.89 lakh hectares (27.70%) of degraded or depleted forest lands. Large area of the landmass in the country including forests have suffered serious degradation. This is more apparent in the case of forests in the country where the degradation is attributed to relentless pressures arising from an ever increasing demand for fuelwood, fodder and timber; and diversion of forest lands to non-forest uses without ensuring compensatory afforestation and essential environmental safeguards, and the tendency to look upon forests as a revenue earning resource.

Maharashtra State which is the third biggest State in the country, is also the third highest in possessing the wasteland. The extent of wasteland in Madhya Pradesh and Rajasthan is 201.42 lakh hectares and 199.34 lakh hectares respectively. In Maharashtra 144.01 lakh hectares of land has been classified as wasteland which comprises 115.6 lakh hectares (i.e. 80.27%) of non-forest and 28.41 lakh hectares (i.e. 19.73%) of degraded forest lands (Table 3.15).

TABLE 3.15: Estimates of Wastelands in India

(ha in lakhs)

States/Union Territories	Non-forest degraded area	Forest degraded area	Total
1	2	3	4
Andhra Pradesh	76.82	37.34	114.16
	(67.29)	(32.71)	(100.00)
Assam	9.35	7.95	17.30
	(54.05)	(45.95)	(100.00)

(Contd.)

TABLE 3.15 *(Contd.)*

1	2	3	4
Bihar	38.96 (71.38)	15.62 (28.62)	54.58 (100.00)
Gujarat	71.53 (91.28)	6.83 (08.72)	78.36 (100.00)
Haryana	29.04 (97.01)	0.74 (02.99)	24.78 (100.00)
Himachal Pradesh	14.24 (72.73)	5.34 (27.27)	19.58 (100.00)
Jammu & Kashmir	5.31 (33.93)	10.34 (66.07)	15.65 (100.00)
Karnataka	71.22 (77.71)	20.43 (22.29)	91.65 (100.00)
Kerala	10.53 (82.33)	2.26 (17.67)	12.79 (100.00)
Madhya Pradesh	129.47 (64.28)	71.95 (35.72)	201.42 (100.00)
Maharashtra	115.60 (80.27)	28.41 (19.73)	144.01 (100.00)
Manipur	0.14 (00.97)	14.24 (99.03)	14.38 (100.00)
Meghalaya	8.15 (42.49)	11.03 (57.51)	19.18 (100.00)
Nagaland	5.08 (36.65)	8.78 (63.35)	13.86 (100.00)
Orissa	31.57 (49.45)	32.27 (50.55)	63.84 (100.00)
Punjab	11.51 (93.58)	0.79 (6.42)	12.30 (100.00)
Rajasthan	180.01 (90.30)	19.33 (9.70)	199.34 (100.00)
Sikkim	1.31 (46.62)	1.50 (53.38)	2.81 (100.00)
Tamil Nadu	33.92 (77.07)	10.09 (22.93)	44.01 (100.00)
Tripura	1.08 (11.10)	8.65 (88.90)	9.73 (100.00)
Uttar Pradesh	66.35 (82.30)	14.26 (17.70)	80.61 (100.00)
West Bengal	21.77 (85.84)	3.59 (14.16)	25.36 (100.00)
Union Territories	8.89 (24.67)	27.15 (75.33)	36.04 (100.00)
Total	936.91 (72.30)	358.89 (27.70)	1,295.80 (100.00)

Notes: 1. Wasteland in Maharashtra increased to 177 lakh hectares in 1990.
2. Figures in brackets indicate percentage
* Barren area notified as forests not included in the above figures.

Source: Society for Promotion of Wasteland Development, New Delhi, 1984

Impact of the Wasteland

Indian forests have suffered deforestation to the extent of over 4 lakh hectares per year for the past so many years. The degradation of forests and other lands has led to serious environmental problems. The result is that the country today faces an acute fuelwood shortage. The country's forests produced 8.39 million tonnes of fuelwood as against an estimated requirement of approximately 114 million tonnes during 1976-77. It means that the forests could meet only 7.00 per cent fuelwood requirement of the country in which majority of population in the country-side still depends on wood as fuel. This shortage was more acute in the states of Karnataka, Assam, Himachal Pradesh Madhya Pradesh and Maharashtra where only 17.45 per cent, 16.18 per cent, 14.62 per cent, 10.00 per cent and 8.84 per cent of the fuelwood requirement was fulfilled by the forests, respectively (Table 3.16). The situation worsened since then. The disastrous results of environmental degradation have begun to appear more severely – the country facing most severe drought of the century today.

The picture is not very different for Maharashtra State. According to the statistics published by the Government of Maharashtra, in the year 1984-85 the State forests produced 389 thousand cu.m of timber and 1,407 thousand cu.m of fuelwood (Table 3.17). This can hardly sustain the pressure of 62 million human population. The pressure of livestock in the State has also increased from 26.36 million cattleheads in the year 1972 to 30.91 million in 1982 (Table 3.18). According to the calculations made by the Forest Development Corporation of Maharashtra, 'the State forests have the carrying capacity of only 39 lakh cattle units, but has to face the incredulous and hopeless situation of supporting 558 lakh cattle units. Which means that only 6.98 per cent of the total cattle units can subsist on forests. (The term cattle unit is different from cattlehead and has been explained in Chapter 4).

Wasteland Development Scheme

With this background in mind, the Government of India has set up Wasteland Development Board (1985). Maharashtra too, has launched an ambitious programme of development of non-forest and forest wastelands of the State. This will be done through afforestation

TABLE 3.16: Fuel Wood Production Scene in India During 1976-77

Sl. No.	States	Population (million)	Total Forest Area (million ha)	Total Produc-topm (million tonnes)	Production per capita (tonnes)	Fuelwood Requirement (million tonnes)	Fuelwood production as % of requirement	Deficit (million tonnes)
1	2	3	4	5	6	7	8	9
1.	Andhra Pradesh	50.000	6.480	0.582	0.011	8.460	6.880	7.878
2.	Assam	50.000	2.860	0.534	0.027	3.300	16.180	2.766
3.	Bihar	67.400	2.930	0.216	0.003	11.220	1.920	11.004
4.	Gujarat	32.600	1.700	0.108	0.003	4.800	2.040	4.692
5.	Haryana	12.300	0.150	0.066	0.005	2.040	3.230	1.974
6.	Himachal Pradesh	4.200	2.170	0.114	0.027	0.780	14.620	0.666
7.	Jammu & Kashmir	5.900	2.070	0.054	0.009	0.960	5.630	0.906
8.	Karnataka	35.200	3.610	1.026	0.029	5.880	17.450	4.854
9.	Kerala	25.700	1.130	0.330	0.128	4.260	7.750	3.930
10.	Madhya Pradesh	52.300	16.830	0.846	0.016	8.460	10.000	7.614
11.	Maharashtra	60.100	6.610	0.880	0.015	9.960	8.840	8.240
12.	Orissa	26.700	6.790	0.246	0.009	4.440	5.540	4.194
13.	Punjab	15.900	0.210	0.024	0.002	2.640	0.900	2.616
14.	Rajasthan	30.300	3.590	0.138	0.005	5.030	2.740	4.892
15.	Tamil Nadu	46.900	2.230	0.174	0.004	7.800	2.230	7.626
16.	Uttar Pradesh	104.500	5.010	1.254	0.012	17.340	7.230	16.086
17.	West Bengal	55.200	1.180	0.318	0.006	9.180	3.460	8.862
18.	Other States	44.100	9.400	0.480	0.011	7.320	6.560	6.840
	(Total India)	720.200	74.950	8.390	0.012	113.870	7.330	102.983

Source: Forest Development Corporation of Maharashtra, Nagpur

TABLE 3.17: Forest Outturn — Major Forest Produce in Maharashtra

(*Qty. in thousand cu.m*)

Year	Timber			Fuelwood		
	F.D.	F.D.C.M.	Total	F.D.	F.D.C.M.	Total
1960-61	332	–	332	1,855	–	1,855
1965-66	416	–	416	2,019	–	2,019
1970-71	540	–	540	1,609	–	1,609
1975-76	294	143	437	874	347	1,221
1980-81	296	165	461	1,148	586	1,734
1984-85	285	104	389	850	557	1,407
1986-87	297	73	370	892	355	1,247
1987-88	280	4	284	576	28	604
1988-89	147	26	173	576	33	909

Source: Maharashtra Forests : A Statistical Outline, Part I — Current Salient Forest Statistics — 1986 Forest Department and Annual Administration Reports, 1990, op. cit.

TABLE 3.18: Livestock in Maharashtra

(*Figures in million*)

Sl. No.	Type	1972	1982
1.	Cows and bullocks	14.70	16.16
2.	Buffaloes	3.30	3.97
3.	Sheep and goats	8.04	10.37
4.	Other livestock	0.32	0.41
	Total	26.36	30.91

Source: (i) Economic Survey of Maharashtra, 1986-87, Directorate of Economics and Statistics, Government of Maharashtra, Bombay.

(ii) Report of the Committee for Afforestation of Barren, Denuded Lands and Hills of Maharashtra, Forest Department, Government of Maharashtra, Nagpur.

and soil-conservation over approximately 2.5 lakh hectares of degraded wasteland in the State annually for next ten years, with the following objectives:

(a) meeting the requirements of fuelwood, fodder, minor forest produce and small timber of the rural populations, and providing means of livelihood to communities traditionally dependent on forests;

(b) increasing the productivity of forests to meet essential national needs;
(c) increasing substantially the forest-cover in the State through massive afforestation and Social Forestry programmes, especially on all denuded, degraded and unproductive lands;
(d) checking soil-erosion and denudation in the catchment areas of rivers, lakes, reservoirs in the interest of soil and water conservation for mitigating floods and droughts and for the retardation of siltation of reservoirs.

The State Level Committee set up by the Government of Maharashtra under the Chairmanship of Shri V.K. Sarwate, presented its report in 1987. The programme suggested by the Committee envisages increasing the present level of afforestation/ plantation from 1.04 lakh hectares annually to approximately 2.5 lakh hectares (Table 3.19).

TABLE 3.19: Annual Afforestation Programmes in Maharashtra Between 1988-89 to 1998-99 on Wasteland

Sl. No.	Agency	Existing target of plantation (in hectares)	Additional target of proposed Plantation (in hectares)	Total (in hectares)
1.	Forest Department	70,000	20,000	90,000
2.	F.D.C.M.	11,000	40,000	51,000
3.	F.L.C.S.	—	40,000	40,000
4.	Social Forestry	23,000	35,000	58,000
5.	Private agency	—	5,000	5,000
	Total	1,04,000	1,40,000	2,44,000

Source: Forest Development Corporation of Maharashtra, Nagpur

The basic objectives laid down by the Wasteland Development Programme are exactly the same as quoted in the National Forest Policy draft of 1987. The basic objectives talk of increasing the productivity of forests to meet 'essential national needs'. Past experience has conclusively shown that the Government forest policy equated national needs with the needs of the commercial, industrial, and State sector.

The policy also mentions 'providing means of livelihood' to forest communities, but the emphasis should be on ensuring the sustenance of forest communities. If the plantations of wastelands are going to be the same as plantations already taken by the Forest Departments,

Forest Development Corporations and Social Forestry, i.e. more than 50.00 per cent of teak and only little of grass and good fodder, fuel species, then there is no meaning in setting high-value objectives in the policy.

B. DIRECT EFFECTS OF THE POLICY

Reduced Availability of Forest-Produce

Fruits, flowers, tubers, roots as food; grass and leaves as fodder; dry-wood, twigs, bamboo as fuel; timber, grass, bamboo and thorns for housing; land for cultivation; herbs, fruits and leaves as medicine; cane, bamboo, leaves, creepers, wild fruits etc. for self-employment and trade; and finally employment in forestry works – all this is provided by forests for those who live in and around the forests. From precious teak to wild grass, every forest produce has utility for tribals. This unique dependence makes tribals inseparable from forests.

Tribals have lived in forests for generations using, though at times even misusing, the natural resource at their wish. However, the fact remains that there never arose the problem of denudation. With the advent of commercial interests, in the form of the British Government, and later continued by the Indian Government, the urban population discovered the use of the natural resource for luxurious purposes. Here lies the origin of tribals' misery. The educated, comfort-loving urban class not only proved to be selfish, which completely ignored the dependence of tribals on forests, but also lacked foresight. Some major policy decisions (already discussed earlier) curtailed the tribals' use of forest produce on the one hand and over exploited the resource to their own (dis)advantage, on the other.

It is very difficult to enumerate the exact extent of reduction in the availability of forest-produce. But a 'pre-investment survey of forest resources' in Chandrapur (Maharashtra) gives an idea of loss of forest produce resulting from clear felling of miscellaneous forest. The study reveals that if approximately one hectare of natural forest is clearfelled, it would result in loss of several species. Out of the total loss (or in other words – out of the total out-turn of timber), broadly speaking, 8.4 per cent loss is of food species, 36.1 per cent is of firewood species, 7.9 per cent of fodder species, 55.3 per cent of construction species and 2.00 per cent of species used as medicines by the tribals. The extent of loss of species which could provide self-employment opportunities to the tribals is as high as 44.6 per cent. (The total sum of per cents exceeds hundred as some of the species can be used for more than one purposes) (Table 3.20).

TABLE 3.20: Species Lost Due to Clearfelling of Miscellaneous Forest (Sample from Chandrapur, Maharashtra)

Sl. No.	Species	Percentage loss	Use of the species
(1)	(2)	(3)	(4)
1.	*Cleistanthus Cellinus*	15.2	Construction timber, firewood.
2.	*Tectona grandis*	13.1	Teak, (seed collection)
3.	*Terminalia crenulata*	7.9	Construction timber, used for agricultural implements making
4.	*Diospyros melanoxylen*	7.5	*Tendu*, *bidi*, Toys, knives handle
5.	*Chloroxylon swietenia*	6.3	Firewood, construction
6.	*Anogeissus latifolia*	4.5	Best source of firewood
7.	*Iagerstroemia parviffora*	4.5	Firewood, fodder, construction, carpentry
8.	*Lannea coromandelice*	3.4	Leaves for fodder/ construction
9.	*Madhuca longifolia var latifolia*	3.2	*Mahua*-country liquor, staple food
10.	*Xylia xylocarpa*	2.9	Construction timber
11.	*Buchanania Lanzan*	2.9	*Charoli*-dry fruit
12.	*Emblica officinalis*	2.3	Aaola-fruit
13.	*Miscellaneous species*	2.0	—
14.	*Dalbergia latifolia*	1.9	*Shisam*, rosewood, expensive timber used for furniture and carving
15.	*Pterocarpus marsupiam*	1.9	Construction, medicine
16.	*Butea monsupium*	1.8	Firewood, leaf-cups, *pattals*
17.	*Acacia Catechu*	1.6	*Katha*, carpentry, agricultural implements
18.	*Cardenia latifolia*	1.6	Firewood
19.	*Nitragyna Parviflora*	1.1	Construction, firewood
20.	*Stereosper mum svacolens*	1.1	Firewood
21.	Rest of the species	13.3	—
	Total	100.00	—

Source: Report on Forest Resources of Wadsa Catchment, Chandrapur, Pre-investment Survey of Forest Resources, Dehra Dun, 1977, Ministry of Agriculture and Irrigation, Government of India.

This data clearly highlight the grave situation for the tribals that has resulted due to the loss of tree-cover on a large scale over several decades. It should further be noted that this loss is a permanent loss of source of food, fuel, fodder and self-employment opportunities for the tribals. With this background, it will be easier to understand the seriousness of the adverse economic impact on standard of living of the tribals.

(i) *Impact on the Standard of Living*

The reduced availability of forest produce had an adverse impact on the living standards of the tribals.

Food

Tribal communities gathered fruits and roots from forests which were an important food item. The fruits differed from region to region and numerous examples can be given, such as – *mahua*, wild bananas, wild mangoes, *charoli, jamun*, dry fruits, wild beries, *bel* fruits etc. The availability of these fruits has gone down in those tracts where forest has been cut and especially where clearfelling has been undertaken. The other reasons for lesser availability of fruits are:

(a) These areas are frequently visited by the outsiders;
(b) Exploitation by contractors who do not exclude even fruit-bearing trees from felling;
(c) Increasing population pressure in rural areas;
(d) Increasing demand from urban areas; and
(e) Forest departments indifference towards plantation of fruit bearing trees.

The tubers and roots, the only source of food for tribals in famines and droughts, are also on the verge of extinction.

Fuelwood

In the earlier days fuelwood was easily available at the door-steps of the tribals. On the contrary, the present situations is that tribal women and children have to spend hours together to collect the fuelwood for everyday use. The hours lost could have been used for other type of gainful employment. It is aptly said, 'Forest famine stares us in face and its dimensions seem frightful. The green revolution may bring us enough food by the year A.D. 2000 but will there be enough fuel to cook it?'

As discussed earlier, the deficit of fuelwood in Maharashtra was 8.24 million tonnes in 1976-77. It has increased to 10.78 million tonnes in the year 1985-86. It is expected to rise further to 14.95 million tonnes by 1990-2000 (Table 3.21).

This gloomy situation is a result of over-exploitation by the contractors and lack of plantations taken up by the forest department for meeting the fuelwood demand. The result is the ever increasing prices of firewood. Between 1960 to 1980 the fuelwood prices increased by 580 per cent in Bombay, 411 per cent in Solapur and 586 per cent in Nagpur (Table 3.22). On an average the increase was by 611 per cent for India for the same period.[18]

TABLE 3.21: Firewood Production Scene in Maharashtra

Sl. No.	Year	Popula-ation (millions)	Total forest area (millions ha)	Total production (million tonnes)	Production per capita (tonnes)	Firewood Requirement (million tonnes)	Production as % of requirment	Deficit (million tonnes)
1.	1976-77	60.100	6.61	0.88	0.015	9.96	8.84	8.24
2.	1980-81	62.780	6.41	1.04	0.017	10.42	9.98	9.38
3.	1983-84	67.040	6.41	0.986	0.015	11.13	8.84	10.144
4.	1985-86	70.030	6.41	0.840	0.012	11.62	7.23	10.780
5.	1989-90	76.430	6.41	0.840	0.011	12.69	6.62	11.850
6.	1999-2000	95.100	6.41	0.840	0.009	15.79	5.32	14.950

Notes: 1. After 1981 Census the population in Maharashtra is expected to have grown at the growth rate of 2.21 per cent per annum.
2. Figures from 1985-86 and onwards are anticipated.
3. Firewood requirement is estimated at 0.166 tonnes/head/annum (Source-National Commission on Agriculture Report 1976, Volume-9, sub-chapter on Production Forestry)

Source : Compiled by Forest Development Corporation of Maharashtra, Nagpur.

Fodder

Cattle ownership is in the culture of tribal-communities and usually they own large number of cattleheads, productive or unproductive. With abundant forest cover tribals never faced any problems with grazing of their cattle. With the dwindling tree cover and increased cattle-heads, came the scarcity of fodder. Between 1972 and 1982 in Maharashtra, the number of cows increased from 14.70 million to 16.16 million, while the number of buffaloes increased from 3.30 million to 3.97 million. The increase in case of sheep and goats together was from 8.04 millions to 10.37 million and from 0.32 million to 0.41 in case of other cattle for the same period, in Maharashtra (Table 3.18). But the capacity of the forest ecosystems of Maharashtra caters to maximum of only 6.98 per cent of the total cattle units subsisting on them in 1982.[19]

TABLE 3.22: Prices of Firewood for 100 kg in Rupees

City	1960	1975	1981	1982	Percentage increase in prices between 1960-81
Bombay	8.5	36.1	57.8	—	580
Solapur	8.9	23.4	45.5	—	411
Nagpur	4.5	12.5	21.0	31.0	586 (on 1982 level)

(For India the increase was by 611 per cent for the same period).

Source: Report of the Committee for Afforestation of Barren, Denuded Lands and Hills in Maharashtra, Forest Department, Government of Maharashtra, Nagpur

Along with the common reasons of increasing population pressure etc., the most important factor is lack of proper management and definite policy of the government. The severeness of the scarcity is not yet being taken up seriously as the plantation of commercially valuable species is still being taken up irrespective of their uselessness as fuel and fodder. The example of eucalyptus is very revealing. Foresters and tree-farmers repeatedly point out that they favour eucalyptus over other species because it cannot be browsed. It is ironic that species which cannot be browsed and are very costly to be used as fuelwood, are planted on large scale.

The impact of the reduced availability of forest produce on the tribals is definitely adverse on their standard of living. It not only results into scarcity of food articles and malnutrition, but also in wastage of working hours in collection of fuelwood and declining quality of the cattle owned by the tribal-communities.

(ii) *Impact on Employment*

Forests are also a major source of employment for the forest-dwelling communities. The forest policy has adversely affected the employment opportunities of the tribals as well.

Reduced Opportunities for Self-Employment -- Tribal communities are well known for their handicrafts made out of the various forest products. These handicrafts are an important source of employment for the tribals. Rope making, basket weaving, cane works, *zadu* (broom) making, pattal making from leaves, bamboo work etc. are some examples of gainful self-employment opportunities for the tribals. With reduced availability of forest-produce, increased restrictions on the entry in forests and rising prices of forest raw material, tribals are finding it very difficult to have adequate and gainful employment for themselves. To what extent the degradation of forests has adversely affected self employment opportunities has already been discussed earlier.

On the other hand, instead of getting the much needed support from the government regarding concessional rates of forest-produce, assured markets, support prices etc., tribals are facing the government's policy which favours the forest-based industries giving away the forest-produce at dirt-cheap rates. Bamboo is a good example to prove the point which has already been discussed earlier in this chapter.

Conflict Between Forest-Operations and Agricultural-Operations

On the one hand the opportunities of self-employment are declining and on the other, the employment generated by the Forest Departments suffers from many limitations. Usually the period of forestry operations, especially plantation, clashes with the agricultural operations. Many tribals are denied the opportunity of earning income through forestry work in this period. According to an official from the Forest Development Corporation of Maharashtra, 15 days around mid-July is a very critical period when agricultural sowing operations are in full swing and the plantations in forests too have to be done in the same period. Therefore, instead of employment local labour, forest departments have to employ outside landless labourers.

Secondly, forestry provides only a partial employment i.e. for three to four months, in a year, and that for limited number of workers except in plantation and felling works. It is obviously, insufficient to provide sustenance to the landless tribals for the whole year.

The need of the time is to plan the forestry works in such a way that they do not clash with the agricultural operations. Special care needs to be taken to provide year-long gainful employment to the forest-dwelling tribals.

(iii) *Displacement of Tribals*

Tribal migration to urban areas is being experienced everywhere in the country.[20] The significant aspect of the mass migration is that it is a forced one and not a voluntary migration. The important reasons for it are:

1. Reduced availability of forest-produce,
2. Lack of employment opportunities, and
3. Displacement due to hydel projects.

The first two aspects have already been discussed and it is obvious that the tribals cannot survive without the forest-produce. Neither can they gainfully employ themselves, nor does the forest department offer them sufficient employment opportunities.

The third reason is significant because in this case it is the government which uproots tribals from forests. A delayed effort (half-hearted) is made to settle them elsewhere in alien surroundings. The tribals are left along to face the social and cultural shock of the changeover. A detailed discussion is taken up here in this regard.

Hydel Projects and Increasing Problems for Tribals

All the major rivers in India, today, are dammed or are in the process of being so. Yet, only a fraction of the country's hydro-power has been realised. In the light of recent experience, questions are now being raised about the viability of large dams which have often proved to be ecological disasters. Silent valley and Munnar in Kerala, Bedthi in Karnataka, Tehri and Vishnuprayag in Uttar Pradesh and Himalayas, Koel-Karo in Bihar, Lalpur in Gujarat, and Bhopalapatnam and Inchampali in Madhya Pradesh, Maharashtra and Andhra Pradesh borders, are a few examples where large scale displacement has taken place or is in the offing. However, the debate on ecological impact of dams has generally remained confined to environmentalists only.[21]

The more important question posed is, whose development do these dams represent and at whose cost? Large dams are generally constructed in areas which have so far not been touched by modern development. The benefits claimed from these projects do not accrue to all the sections of the population. The irrigation benefits those who have lands down the dams. Tribals have neither the large holdings to be able to maximise the use of irrigation facilities nor the capacity to pay for them. Their lands generally come in the catchment areas. On the other hand, they face several problems which can be listed as follows:

1. Tribal life and culture is unequally interwoven and integrated with the forests. Once uprooted, tribals are totally disoriented.
2. Compensation for land and houses is usually grossly inadequate.
3. No compensation is paid for the loss of employment and other sources of livelihood.
4. Landless labourers, marginal farmers, sharecroppers, tenant-cultivators and those who live on collecting and selling minor forest produce have to suffer the most.
5. Small farmers who supplement their income by rearing cattle are hit because new settlements invariably have inadequate grazing lands. They have to sell off their cattle at throwaway prices before shifting.
6. People engaged in non-agricultural activities like fishing etc. are not entitled for compensation.
7. A major part of compensation has to be spent on shifting to new site.
8. When only a part of a village is shifted, the families left behind face isolation.
9. Even in terms of employment generated during the period of construction, usually skilled labour is brought from outside

and local labour is absorbed only partially in casual and temporary employment.

> "This means a sudden transition from a settled life in relative isolation to a marginalised life dominated by outsiders."[22]

10. The settlements done are in distant places, sometimes in a totally alien environment and culture, thus creating insurmountable adjustment problems.

Many times no efforts are made even to resettle the displaced people. To boost the cost-benefit ratio for obtaining Planning Commission's approval and finance from international agencies like the World Bank, forest loss is grossly undervalued on the one hand and in order to reduce the rehabilitation expenditure on displaced persons, the tribals are simply evacuated, on the other.

'Land for land' is a better policy than cash settlements, but this land is never in the command area. In most cases either the forest-land cleared or waste fallow land is allotted without any provision for developing the land or for the supply of necessary inputs. Maharashtra is the only State in India which has a legislation that recognises the right of people displaced by irrigation, power and other public utility projects to get land according to the 'Maharashtra Resettlement of Project Displaced Persons Act, 1976'. However, this measure alone has not helped solving the whole problem of tribal displacement.

Summary

It can be surmised from the above discussion that the Indian Government has always looked at forest wealth as a source of revenue.

The economically powerful commercial class has succeeded in pressurising the government to give first order priority to the industrial needs of the forest produce.

The elite group of ecologists, under the academically powerful leadership of scholars, has conveyed its concern over the degradation of forests. Yet, the government has failed to achieve the essential target of converting 33.33 per cent of geographical area into forest cover.

It is only the illiterate, poor tribals — the actual suffers, who are unable to convey with clarity and seriousness, the adverse effects of deforestation on themselves. Clear felling by Forest Development Corporations, eucalyptus and teak plantation under Social Forestry, the present rehabilitation policy, stringent measures by reducing tribal-concessions to save forests etc. are clearly anti-poor and anti-tribal.

NOTES

1. Report of the Scheduled Areas and Scheduled Tribes Commission, 1960-61, Government of India, Manager of Publications, New Delhi, pp. 127-128.
2. *The State of India's Environment, 1982*, A Citizens Report, Centre for Science and Environment, New Delhi, p. 34.
3. *Ibid.*, p. 34.
4. Roy Burman, B.K., 'Issues in Environmental Management Centering Forest and Role of Tribal Communities', *South Asian Anthropologist*, 1985, pp. 41-42.
5. Devi, Mahashveta, 'Eucalyptus : Why?', *Economic and Political Weekly*, vol. XVIII, No. 32, August 6, 1983, p. 1380.
6. *The State of India's Environment*, 1984, A Second Citizen's Report, Centre for Science and Environment, New Delhi, p. 74.
7. *Ibid.*, p. 76.
8. Gadgil, M. *et al.*, 'Forest Management and Forest Policy in India : A Critical Review', *Forests, Environment and People*, (1984), p. 16.
9. Sharma, A., 'Forestry in India's Economy : New Approach to Forest Management', Forest Development Corporation of Maharashtra, Nagpur, p. 3.
10. *Ibid.*, p. 5.
11. Chowdhary, K., 'Managing Forests for Development', Report No. 88, Centre for Science and Environment, New Delhi.
12. Kulkarni S.D., 'Social Forestry : Need for a Proper Perspective and Policy', *Artha Vijnaa*, Sept. 1983, V. 25 N.3, p. 290.
13. Project Paper, Social Forestry Project, Maharashtra State, Government of Maharashtra.
14. Chowdhary, K., 'Managing Forests for Development', *op. cit.* p. 5.
15. *Ibid.*, p. 4
16. 'Anti Social Forestry', Report from *Financial Express*, Wed., July 13, 1983.
17. Report from *Lokmat* (a Marathi Daily from Nagpur), Nov. 1 and 2, 1987, Lokmat Printing Press, Nagpur.
18. Agarwal, V.P. *et al.*, 'Energy From Trees: Potentials and Problems of a Second Green Revolution in India,' Report of the Technical Sub-Committee of the Proposed Wasteland Development Society, (1981), Government of India, New Delhi, p. II-3.
19. (i) Report of the Committee for Afforestation of Barren, Denuded Lands and Hills of Maharashtra, (1982), Government of Maharashtra.
    (ii) Economic Survey of Maharashtra, 1986-87, Directorate of Economics and Statistics, Government of Maharashtra.
20. 'Anti-People Development — Case of Inchampali Project', *Economic and Political Weekly*, vol. XX No. 22, June 1, 1985, p. 953.
21. The State of India's Environment, 1984, *op. cit.* p. 100.
22. 'Anti-People Development', *op. cit.* p. 952.

PART II

4

FORESTS IN MAHARASHTRA AND SELECTION OF SAMPLES

This chapter mainly deals with the methodology adopted for the selection of sample households from the four administrative circles of the Forest Department of Maharashtra. Before that, a brief description of the geo-climatic nature of the forests in the State, and the tribal population residing in these forests has been presented. A brief discussion on the choice of the three types of samples that have been used, is also incorporated.

Geo-Climatic Features of Maharashtra

Maharashtra State has a fairly large area of which nearly 60,955 sq. km area is classified as forests. These forests were, once, under very good sylvan cover. The forests are spread over strategic watersheds, low mountainous belts of Satpuras (in north), Sahyadris (in west) and local hill ranges of Ajantha, Satmalas (in centre) etc. They encompass a wide climatic range, from semi-arid to very humid; rainfall varying from 250 mm in Junnar division to 6500 mm at Amboli in the Western Ghats. They also represent a wide topographical variation from extensive coastal strip in the west, the Deccan Plateau, the Eastern plains, and hills rising 1178 metres in Satpuras in Melghat to 1650 metres at Kalsubai in Western Ghats in Nasik division.

The areas cover wide variation of edaphic conditions; soils varying from deep clayey black cotton soils, gritty, bouldery soils with soil depth of a few c.m. with rocky outcrops or sheet rock near the surface, to almost pure sandy soils.[1]

The forests naturally represent a corresponding wide variation from semi-arid scrub to comparatively stunted semi-evergreen forests. The state is also reputed for some excellent teak and mixed forests in Chandrapur, Melghat, Thane and Dhule.

At present, though 20 per cent of the geographical area is termed as forests, large portion is degraded, comprising barren hill slopes and lands. There are no virgin forests in the state, and large forest areas are under-stocked.

Tribal Areas and Tribal Population

The tribal areas of Maharashtra are not concentrated in one compact contiguous zone, but are spread over in three distinct regions, namely, the western (the Sahyadri), the northern (the Satpuda) and the eastern (the Gondvana) belts. Majority of the tribals in the State reside in the Sahyadri belt which covers parts of Thane, Raigad, Pune, Ahmadnagar, Nasik and Dhule districts. The Satpuda belt covers northern parts of Dhule, Jalgaon and Amravati districts, whereas the Gondvana belt comprises areas from Chandrapur, Nagpur, Bhandara, Yavatmal and Nanded districts. These three distinct regions comprise 57.72 lakhs of scheduled tribe population which constitute 9.19 per cent of the State's population according to the 1981 census. Quite sizeable tribal areas are still inaccessible for about five to six months during a year.[2]

Government of India have declared predominant tribal areas as scheduled tribal areas, and the tribal population is separately enumerated for each district in Maharashtra. However, in the 1961 Census, scheduled tribe population was enumerated only for 21 districts excluding Buldhana, Akola, Wardha, Nagpur and Bhandara districts. It was only in 1981 census, for the first time, that the tribal population was separately enumerated for all the districts of Maharashtra. Total tribal population in Maharashtra increased from 23,97,159 in 1961 (i.e. 6.06 per cent of the total population of the state) to 57,72,038 in 1981 (i.e. 9.19%).

The Revenue Department of Maharashtra has divided the State into three regions, namely, Western Maharashtra, Marathwada, and Vidarbha; and six divisions, namely, Konkan, Nasik, Pune, Aurangabad, Amravati, and Nagpur. It is interesting to note that the divisions with high percentage of tribal population are also the divisions with large forest area controlled by the Forest Department. The divisions with high tribal concentration are Nasik, Amravati, Nagpur and Konkan with respectively, 33.54 per cent 13.73 per cent, 23.26 per cent and 17.96 per cent of the state's tribal population according to 1981 Census (Table 4.1). In Maharashtra the forest area under forest department is 56,133.66 sq.km according to 1980-81 figures which has only marginally declined from 56,190.38 sq.km in 1970-71. According to the 1980-81 figures, 10,332.40 sq.km (18.40%) of forest area is in Nasik division, 8,522.26 sq.km (15.18%) in Amravati division, 23,571.16 sq.km (41.99%) in Nagpur division and 5,588.57 sq.km (9.95%) in Konkan division. It is only in Nasik division that the forest area has increased by 489.66 sq.km between 1970-71 and 1980-81 (because of the acquisition of private forests in Dhule during the decade). In 1988-89 area under Forest Department declined to 55,900.29 sq.km (Table 4.2).

The field survey for the sample of 'households from forest villages', was done from the four administrative circles of the Forest Department which are in the above-mentioned four Revenue divisions. The circles selected were Amravati, Dhule, Thane, and South Chandrapur — these have high proportion of tribal population as well as of forest area. The other details about sample selection are given later in this chapter.

The main tribes in Thane circle are — Varli, Koli Malhar, Thakur, Kothodi, Kokna, Mahadeo Koli, Agris, Katkari, Dublas and Dhodiahs.

Dhule circle tribals mainly belong to Bhil, Dhanka, Gamit, Kokna, Koli, Dhor, Pardhi, Naikda and Kathodi tribes.

South Chandrapur has Gond, Madia, Bhil, Pradhan, Halba and Pardhi tribes.

Amravati circle has tribals belonging mainly to Korku tribe.

Forest Areas, Administration, Rights and Privileges

Presently Maharashtra has 60,955 sq.km of forest area which is 19.80 per cent of the State's total geographical area. Out of the total forest area, 55,900 sq.km (i.e. 91.70%) is under the administration of Forest Department and 5,055 sq.km (i.e. 8.30%) is under Revenue Department according to 1988-89 figures. The area under Forest Department is classified as Reserved forests (38,491 sq.km i.e. 68.85 per cent of area under Forest Department), protected forests 15,122 sq.km i.e. 27.05%), and unclassed forests (2,286 sq.km i.e. 4.08%). The total forest area under Revenue Department is also similarly classified as Reserved forests (1416 sq.km i.e. 28.04 per cent of forest area under Revenue Department), protected forests (137 sq.km i.e. 2.71%), and unclassed forests (3,502 sq.km i.e. 69.27%).

TABLE 4.1: Region and Divisionwise Tribal Population in Maharashtra, 1960-61 and 1980-81

Sl. No.	Region/Division	1960-61			1980-81		
		Tribal Population	% to tribal population of Maharashtra	% to total population of Maharashtra	Tribal population	% to tribal population in Maharashtra	% to total population of Maharashtra
1.	**Western Maharashtra**	19,00,588	79.28	4.80	32,37,195	56.08	5.15
	Konkan Division	6,21,468	25.92	1.57	10,37,054	17.96	1.65
	Nasik Division	11,75,588	49.04	2.97	19,36,376	33.54	3.08
	Pune Division	1,03,532	4.31	0.26	2,63,765	4.56	0.42
2.	**Marathwada**	102,650	4.28	0.25	3,98,718	6.90	0.63
	Aurangabad Division	102,650	4.28	0.25	3,98,718	6.90	0.63
3.	**Vidarbha**	3,93,921	16.43	0.99	21,36,125	37.60	3.40
	Amravati Division	2,10,490	8.78	0.53	7,93,036	13.73	1.26
	Nagpur Division	1,83,431	7.65	0.46	13,43,089	23.26	2.13
	Total Tribal Population in Maharashtra	23,97,159	100.00	6.06	57,72,038	100.00	9.19
	Total Population of Maharashtra	395,54,000	—	—	6,27,84,000	—	—

Source: Statistical Abstract of Maharashtra, Directorate of Economics and Statistics, Government of Maharashtra, Bombay.

TABLE 4.2: Region and Divisionwise Forest Area Under Forest Department in Maharashtra, 1970-71, 1980-81 and 1988-89

(Area in sq.km)

Sl. No.	Region/Division	1970-71	1980-81	1988-89
1.	**Western Maharashtra**	20,234.59 (36.01)	21.238.75 (37.83)	21082.30 (37.71)
	Konkan Division	5,786.80 (10.29)	5,588.57 (9.95)	5,572.34 (9.96)
	Nasik Division	9,842.74 (17.51)	10,332.40 (18.40)	10,050.47 (17.98)
	Pune Division	4,605.05 (8.19)	5,317.78 (9.47)	5459.50 (9.77)
2.	**Marathwada**	2.815.25 (5.01)	2,801.49 (4.99)	2,818.61 (5.04)
	Aurangabad Division	2.815.25 (5.01)	2,801.49 (4.99)	2818.61 (5.04)
3.	**Vidarbha**	33.140.54 (58.97)	32,093.42 (57.17)	31,999.37 (57.24)
	Amravati Division	8,571.50 (15.25)	8,522.26 (15.18)	8,350.21 (14.93)
	Nagpur Division	24,569.04 (43.72)	23,571.16 (41.99)	23,649.16 (42.30)
	Total for Maharashtra	56,190.38 (100.00)	56,133.66 (100.00)	55900.29 (100.00)

Note: Figure in brackets denote percentage.

Source: Forest Statistics of Maharashtra, Forest Department, Government of Maharashtra, Pune.

It is important to note that while the total forest area in the State has only marginally declined from 63,469 sq.km in 1960-61 to 60,955 sq.km in 1988-89, the extent of decline in the forest area under Revenue Department is more (i.e. by 23.47%) than the decline in area under Forest Department (i.e. by 1.73%). The decline in forest area under Revenue Department is mainly in case of Reserved forests (from 3,126 sq.km in 1960-61 to 1,416 sq.km in 1988-89 i.e. by 54.70%). The percentage decline in case of protected forests under Revenue Department is 28.73 while the unclassed forest area under the department has increased by 6.50 per cent for the same period.

On the other hand the area of Reserved forests under Forest Department has increased from 36,532 sq.km in 1960-61 to 38,491 sq.km in 1985-86 i.e. by 5.36 per cent. The percentage decline in case of protected and unclassed forests under Forest Department is 12.09 and 27.54 respectively for the same period (Table 4.3).

TABLE 4.3: Legal Classification of Forest Area in Maharashtra under Forest and Revenue Departments

(*Area in sq.km*)

Sl. No.	Class of Forests	Department in charge	1960-61	1970-71	1980-81	1988-89
1.	Reserved	F.D.	36,532 (64.22)	37,774 (66.36)	38,455 (68.51)	38,491 (68.85)
		R.D.	3,126 (47.32)	2,819 (45.76)	1,458 (28.37)	1,416 (28.01)
2.	Protected	F.D.	17,203 (30.24)	16,620 (29.20)	15,172 (27.02)	15,122 (27.05)
		R.D.	192 (2.91)	187 (3.04)	153 (2.98)	136 (2.70)
3.	Unclassed	F.D.	3.155 (5.54)	2,526 (4.44)	2,507 (4.47)	2,286 (4.09)
		R.D.	3,288 (49.77)	3,154 (51.20)	3,528 (68.65)	3,501 (69.27)
	Total	F.D	56,890 (100.00)	56,920 (100.00)	56,134 (100.00)	55,900 (100.00)
		R.D.	6,606 (100.00)	6,160 (100.00)	5,139 (100.00)	5055 (100.00)
	Grand Total		63,496	63,080	61,273	60,955

Note: F.D. is Forest Department and R.D. is Revenue Department
Source: 'Forest Statistics of Maharashtra', *op. cit.*

The important reasons for decline in forest area under both the departments are — regularisation of encroachments, submergence under river valley or other irrigation projects, rehabilitation of project affected persons, township housing, roads, railways etc. Some forest area was also transferred from Revenue Department to Forest Department for the purpose of afforestation.

In the present study the area under Forest Department only, has been considered. For the administrative working of the department, the forest area of the State is divided into eleven circles, namely, Thane, Nasik, Pune, Kolhapur, Aurangabad, Amravati, Dhule,

Nagpur, Yavatmal, South Chandrapur, and North Chandrapur. Each circle is further divided into 'divisions', divisions into 'ranges', and ranges into 'beats'.

According to 1988-89 figures, 18.48 per cent of the area under Forest Department is in South Chandrapur circle, 9.04 per cent in Thane circle, 9.80 per cent in Dhule circle and 5.63 per cent in Amravati circle (Table 4.4).

TABLE 4.4: Circlewise Area under Forest Department in Maharashtra

(Area in sq.km)

Sl. No.	Name of the Circle	1960-61	1980-81	1985-86	1988-89
(1)	(2)	(3)	(4)	(5)	(6)
1.	Thane	5.841 (10.25)	5,210 (9.28)	5,108 (9.11)	5,054 (9.04)
2.	Nasik	9,567 (16.80)	8,842 (15.75)	4,594 (8.20)	4,566 (8.16)
3.	Dhule	–	–	5,587 (9.97)	5,482* (9.80)
4.	Pune	7,659 (13.45)	3,329 (5.93)	1,838 (3.28)	1,850 (3.50)
5.	Kolhapur	–	3,857 (6.87)	4,033 (7.20)	4,040 (7.22)
6.	Aurangabad	–	2,800 (4.99)	2.803 (5.00)	2,816 (5.03)
7.	Amravati	8,887 (15.6)	8,524 (15.18)	3,319 (5.92)	3,148 (5.63)
8.	Yavatmal	–	–	4,893 (8.73)	4,894 (8.75)
9.	Nagpur	6.683 (11.75)	6,246 (11.13)	6,284 (11.21)	6,322 (11.31)
10.	South Chandrapur	18,309 (32.15)	17,326 (30.87)	10,337 (18.45)	10,330 (18.48)
11.	North Chandrapur	–	–	6,800 (12.14)	6,861 (12.27)
12.	Tadoba National Park	–	–	134 (0.24)	134 (0.24)
13.	Tiger Project	–	–	308 (0.55)	308 (0.55)
	Total	56.949 (100.00)	56,134 (100.00)	56,045 (100.00)	55,900 (100.00)

Note: Total of 1988-89 includes 87.85 sq.km are of Borivali National Park.

Source: Forest Statistics of Maharashtra, *op. cit.*

Rights and Privileges of Forest-Dwellers

As already mentioned, in Maharashtra, 9.19 per cent of the total population is tribal according to 1981 Census. More than 90.00 per cent of this tribal population lives in forests and depends on forests for its basic needs.[3] Before the advent of the British, the tribal dependence on forests was total and unrestricted. It was the British intrusion which disturbed the symbiotic relation between the forests and the tribals, converting the tribals from forest owners to mere 'concession and privilege holders' in forests.

The *Bombay Forest Manual*, Vol. III, which deals with the rights and concessions of the tribals, begins its chapter on 'Forest Privileges' with a bold announcement that, 'in its management of forests, the ease and contentment of the people is an object of greater solicitude to Government than the realization of revenue',[4] and also that, 'Government have no desire to increase their forest revenue by the curtailment of conceded privileges or of local supply or by the levy of excessive charges for grass and other minor forest produce'.[5] It further adds that, 'The benefit of any revenue so obtained would be altogether insufficient to countervail the hardship and irritation that would be caused were the rayats unduly pressed in the matter of obtaining 'rab' and firewood, grazing for their cattle or grass for thatching and other household purposes'.[6]

But the above mentioned grant of privileges is immediately followed by details on 'powers of offices to withdraw and suspend privileges, privileges only for definite periods, and how the government shall abandon no part of their proprietary rights by granting privileges. It is also made clear that these privileges are granted as a matter of favour and not as a right, when the manual states that, 'Government have always made it a point while formally conceding liberal privileges on the completion of forest settlement to communities which are reported to have enjoyed them, though as a matter of favour and not of right, for many years past, to make it clearly understood that such privileges are intended to be exercised as of favour only and are subject to withdrawal'.[7]

Though the concessions and privileges differ from circle to circle according to the availability, the general privileges for all forest areas in Maharashtra, as given in the Manual, are as follows:

General Privileges

(1) No one will be prohibited from drawing water obtainable in forest in cases where it is not procurable elsewhere within a reasonable distance.

(2) (a) No change will be made under section 55, Land Revenue Code, for the provision of water for agricultural purposes from any stream, budki, bandhara, tank and all natural sources of water in forest areas on which no expenditure has been incurred by Government at any time.

(b) No fee will be levied when permission is granted for digging wells or channels in forest areas for agricultural purposes.

(c) Permission will be given freely for the clearance of choked up tanks and channels, and for the removal of any forest growth obstructing the flow of water.

(3) Villagers having right to way to water through forest are entitled to a path 50 feet wide which they will be allowed to fence with thorns obtainable free of charge on application to the Divisional Forest Officer (DFO). The villagers are also allowed to keep such paths free of all undergrowth.

(4) Free grazing in open forests under passes issued by the Revenue Department, for village cattle (including sheep and goats where permitted) of forest and non-forest villages and non-village cattle i.e. both cattle of professional graziers and cattle which do no remain in one village. This concession is subject to restrictions as regards admission as laid down in the grazing rules in force.

(5) No forest will be closed to grazing within a quarter of a mile of a village site.

(6) Access to the grazing areas in the interior will be granted by the allotment of sufficiently wide short-cut approach roads. The short-cut cattle paths leading through closed forest to open forest will be marked by the Forest Department in order to facilitate fencing by the villagers in the manner indicated in paragraph 3 above.

(7) Removal of stones and earth, from places approved by the DFO, for the gatherers' own domestic or agricultural use.

(8) Removal of fallen leaves and grass for the gatherers' own bonafide agricultural and domestic uses.

(9) Removal from coupes under exploitation, before, commencement of plantation operations therein,

(a) of felled timber not useful to and therefore left by contractors, and

(b) of branchwood of felled trees of the size stipulated in contract agreements, for the gatherers' own domestic or agricultural use.

(10) In case of destruction of houses by fire in forest areas timber of inferior species required for temporary huts will be made available with the utmost promptitude by the Range Forest Officer on production of a certificate from the Mamaltdar or Mahalkari concerned, up to a value of Rs. 10. The timber of only species as have been prescribed by the conservator will be granted.

All these general privileges are granted in all the eleven circles with some modifications and specifications regarding quantity etc. The first three general privileges concerning 'water' have not been modified. Other changes made are given below for the four circles selected for field study.

Grazing

A. *Amravati Circle*

According to Grazing rules of 1952-53, four units of cattle can graze free of charge. Other cattle units get concessions in proportion to the land owned by the privilege holder. If the grazing land allotted is far off from the village, grazing fees is reduced by 25.00 per cent. Regular cattle can graze in the Reserved forests daily, but stray cattle are not allowed at all.

B. *Dhule Circle*

There is no special mention about any extra facility. Instead, 'with the exception of the rights in *Bombay Forest Manual*, Vol. III, para 93, the forests are not burdened with any prescriptive rights.'

C. *Thane Circle*

In this circle also there is no specific mention regarding grazing privilege but for, '. . . no privileges are recognised, but grass, seed and leaves may be removed with the permission of the Divisional Forest Officer (DFO),' Removal of grass is permitted for *'rab'* (i.e. burning of grass in the agricultural fields before sowing) purposes.

D. *South Chandrapur Circle*

In this circle the villagers have been classified as cultivators and non-cultivators. Cattleheads owned by the cultivators are further classified as — useful for cultivation and otherwise. The grazing fee is of two types — concessional and commercial.

Before specifying the concessions, the 'unit' of cattle is explained:

Cow or bullock (1 to 3 years)	– 1 unit
Buffalo (2-3 years)	– 2 units
Cow calf	– 1/2 unit
Buffalo calf	– 1 unit
Up to 5 calves	– 0 unit

The grazing concessions are as follows:

(a) cattleheads used for cultivation —

(i) cultivators have up to 8 units of cattleheads and
(ii) cultivators having more than 8 units get concessions only for 6 units of cattlehead per plough.

(b) Concessions —

(i) cultivators, agricultural labourers, carpenters, black-smiths etc. can graze four units of cattle free of charge. And up to four extra units of cattle can be grazed at concessional rates.
(ii) Those who have more than eight units of cattle can graze six cattle units per plough at concessional rates. But others are charged at commercial rates.

Grass and Fallen Leaves

A. *Amravati Circle*

Grass, thorns can be fetched from the coupes selected for felling within 8 km of every village. This can also be sold by the villagers.

B. *Dhule Circle*

Grass and fallen leaves can be taken even from Reserved forests. Amount of grass or leaves that can be collected has not been specified.

C. *Thane Circle*

(1) Grass, reeds and dead leaves can be collected and used by the inhabitants of forest villages for rab purposes.
(2) Fourteen shrubs, namely, *Karvanda, Babli, Kala-Kuda, Ukshi, Bhar-Jambhul, Mastodi, Kuda, Gomati, Torni, Bhokar, Kuda* and *Dhadti* and other branchwood of no value can be cut for woodash, manure and rab purposes by the inhabitants of forest villages.
(3) Removal of matcrial felled or lopped-off which is less than six inches in circumference, grass, reed, leaves etc., except bamboo and palms from coupes under exploitation, is allowed. Material left by contractors in the coupes is allowed to be removed from each sub-coupe before the date on which preparation of the ground for plantation begins in such sub-coupes. If the coupes of the given year, in a block, remain unsold, special arrangements are made for allowing the people to enter such coupes and cut the material, under supervision.

D. *South Chandrapur Circle*

Within 8 km of any village, thorns and shrubs for fencing are available on the rates fixed by the government. Green shrubs cannot be removed.

Branchwood and Timber

A. *Amravati Circle*

All privileges which were enjoyed by people when residing in ex-proprietory forest villagers, are continued. Timber with 60 cm girth can be used for domestic purposes at concessional rates from the coupes selected for felling.

B. *Dhule Circle*

For domestic use and for making agricultural implements, timber of any species dried and thrown away (i.e. of no other use) can be taken. There is, however, restriction on the quantity, namely, as much as can be carried on head, and not more.

C. *Thane Circle*

Fallen dead timber and firewood of species other than teak (*Tectona grandis*), tiwas (*Ougenia albergioides*), Khair (*Acacia Catechu*), Shisham (*Dalbergia Latifolia*), Ain (*Terminallia torrentosa*), and sandal-wood (*Santolum Album*) can be taken by the inhabitants of forest villages and by hill tribes for home consumption only. Following 14 hill and other tribes in the district, except North Salesetta peta, can take it for sale or barter. The tribes are — Bhils, Dhodias, Dublas, Koknis, Katkaris, Malharkolis, Phaseparadhis, Kaikaris, Ramoahis, Thakurs, Vadars, Vaghris, Vaitis, and Varlis.

D. *South Chandrapur Circle*

(1) Any type of timber below 24" in girth at breast height over bark, except of teak, *shisham, bija, tinsa, haldu* and edible fruit bearing trees like *mahua,* mango, *harra, achar* etc. can be taken.

(2) The source of timber —

  (i) through *Tahsildar* — timber worth up to Rs. 200.
  (ii) through Deputy Collector — timber worth up to Rs. 500.
  (iii) through Collector — timber worth above Rs. 500.

(3) Teak is not available free of cost, but any villager can buy it by applying to Revenue Officer.
(4) Necessary amount of timber required for agricultural implements is freely available. It is also available for sale against a pass, at market rate between 15th September to 15th November, and also in June and July.

Firewood

A. *Amravati Circle*

(1) Any timber up to 24" girth can be obtained from leased forest area of coupe being worked.
(2) Wherever there is scarcity of firewood, matured and useless trees can be sold at some concessional rates.
(3) Dry firewood up to 20 cartloads can be collected by the inhabitants.

B. *Dhule Circle*

Nothing special is mentioned in addition to privileges under *Bombay Forest Manual*, Vol. III.

C. *Thane Circle*

Same as mentioned in the previous heading regarding timber and branchwood.

D. *South Chandrapur Circle*

(1) Any timber (except some species) less than 24" in girth can be removed from forests at half the market price.
(2) If girth is more than 24", full market price is to be paid.
(3) Within 8 km of a village any kind of timber can be removed at rates fixed by the government.
(4) Any unsold or left-over timber can be taken by the tribals.
(5) Timber from working coupes can be used by the tribals.

Bamboo

A. *Amravati Circle*

(1) Bamboo is given at concessional rate of Re. 1.25 per 100 bamboos, for only the specified 30 villages of the Amravati circle.
(2) From working coupes, bamboo can be taken out anytime for domestic consumption only. For sale, it can be taken out in the period between 1st January to 30th April.
(3) For the residents of Chikhaldara, 200 number of dry bamboos can be taken for construction of huts, while 100 number of bamboos can be taken for repairing huts, every year. Green bamboos, only up to 50 pieces, can be brought at concessional rate.
(4) For Burud tribe, the basket weavers, 1,500 bamboos can be contained against a sale-pass at concessional rates. The rates are revised time and again, according to the availability.

B. *Dhule Circle*

No mention is made in this regard.

C. *Thane Circle*

Bamboo and *Karvi (Strobianthus callous)* can be taken out by inhabitants of forest villages, and hill tribes, within the limits of the forest block in which their villages are situated, for bonafide local consumption. They may cut and remove *Karvi* as headload for sale outside their villages.

(a) Bamboo removed for cottage industries in Dahanu Division is charged at the rate fixed for non-commercial purposes.
(b) The aboriginal and hill tribes in the district are permitted to prepare and export, once a week, bamboo, bamboo articles worth Rs. 3.50 per head, free of charge and without being asked for passes.

D. *South Chandrapur Circle*

(1) Dry bamboo can be taken free of cost for domestic use.

(2) Green bamboo can be taken at half the market rate for domestic use.

(3) For Burud or Basod people, 1,500 green bamboos are sanctioned for their cottage industry, at concessional rates, per family per year.

Minor Forest Produce

A. *Amravati Circle*

(1) *Mahua* flowers and fruits, *charoli*, gum and other edible items can be collected from reserved forests free at cost.

(2) For tanning purposes cobblers get *Babhul* at concessional rates.

(3) Blacksmiths can get coal at concessional rates from the coupes felled.

(4) Carpenters can get teak wood at concessional rates.

(5) Residents of Chikhaldara can get teak leaves.

B. *Dhule Circle*

Mango, Tamarind, *Umber, Amla, Karwanda* and *Jamun*, etc. fruits can be taken by the villagers.

C. *Thane Circle*

Minor Forest produce other than Myrabolams (*Terminalia Chebula*), *Beheda* (*Terminalia beleria*), nuts, fruits of toddy-palm, pods of *shikekai*, and such articles as may be specially reserved for exploitation from year to year, can be taken. Other products that can be taken by the inhabitants of forest villages and hill tribes, for domestic or agricultural use or for sale, are —

(a) *Apta, temburni* leaves, leaves of *bel*, flowers, fruits of *Mahua*, and *Chillhari* bark — for personal use only. Forest villagers may remove mahua fruits for sale also;

(b) Edible tubers and roots, jungle fruits, flowers, honey, wase, gum and medicinal plants can be taken for domestic use as well as for sale;

(c) Leaves of Toddy-palm can be used for home consumption

only by the agriculturists of forest villages; and

(d) Thorny shrubs and leaves of palas can be used by hill tribes for domestic use or for sale, and by the inhabitants of forest villages for their own use only.

D. *South Chandrapur Circle*

(1) Fencing material, grass, creepers, roots, leaves, flowers and fruits can be taken by the villagers free at cost.

(2) *Mahua* fruits and mangoes, *achar* fruits etc. can be used without payment of any kind.[8]

After satisfying the bona fide needs of the villagers, the villagers outside the nistar zone can avail of the above-mentioned facilities on payment of the charges as fixed by the Collector for each of the item.

The Reality

The list of privileges and concessions enumerated above may seem to be sufficient to meet the needs of the tribals. But, in reality, it is not so. Many times tribals are not able to make use of these privileges because of the non-availability of forest produce. If a study of the use of forest produce is made, one finds that only marginal use of the forest produce falls to the lot of these privilege-holders. Major portion of the forest produce is collected by the Forest Department and private purchasers — usually the traders.

These three agencies namely 'Forest Department', 'Purchasers', or Forest Labour Cooperative Societies and 'Right and Privilege-holders and Free-grantees', mainly exploit the forest produce. The forest produce is classified in three categories, namely timber, fuelwood and minor forest produce. The available data convincingly show that it is always the Forest Department that has the highest share in the exploitation of forest produce, and it is followed by the share of purchasers. In 1971-72 Forest Department exploited 58.32 per cent of the total timber exploited and the purchasers had 40.45 per cent share of it. The share of Forest Department increased to 68.15 per cent in 1987-88. After 1982, 'Purchasers' were replaced by the Forest Labour Cooperative Societies (FLCS) which had a share of 36.72 per cent of the timber in 1984-85. The share of the privilege-holders remained a paltry 1.21 per cent in 1971-72 and 1.76 per cent in 1984-85, which further decreased to 0.45 per cent in 1987-88.

In case of fuelwood, 'purchasers' had the highest share in its exploitation (i.e. 59%) in 1971-72. It only marginally reduced to 46.54 per cent in 1984-85 — exploited by the FLCS. On the other hand, the share in fuelwood exploitation of the Forest Department increased from 36.62 per cent to 41.58 per cent over the same period. Rights and privilege-holders had only 4.36 per cent of their share in 1971-72 which increased to 11.86 per cent in 1984-85. It reduced to 6.52 per cent in 1987-88. But it is important to note here that though there is a percentage increase in the share of Rights and Privilege-holders between 1971-72 and 1984-85 there is only marginal increase in the quantity of fuelwood extracted by them. In 1971-72 the Privilege-holder had 74.385 thousand cu.mt. of fuelwood which increased to only 101.126 thousand cu.mt. The total production of fuelwood in this period decreased from 1705.316 thousand cu.mt. to 851.956 thousand cu.mt.

Minor Forest Produce also was exploited mainly by the department to the extent of 76.68 per cent in 1971-72 which declined to 49.24 per cent in 1984-85 but rose substantially to 76.05 per cent in 1987-88. The share of 'purchasers/FLCS' increased from 22.44 per cent in 1971-72 to 49.47 per cent in 1984-85, but decreased to 22.45 per cent in 1987-88. In the case of Privilege-holders, it increased only marginally, from 0.86 per cent to 1.28 per cent between 1971-72 and 1984-85.

The figures of the total production of the forest produces mentioned above, also give an interesting picture. The total production of timber has gone down from 430.364 thousand cu.mt. in 1971-72 to 280.278 thousand cu.mt. in 1987-88. Similarly, in case of fuelwood, the total extraction in 1971-72 was 1705.316 thousand cu.mt. which reduced to 576.582 thousand cu.mt. in 1987-88. Though the MFP extraction which was worth Rs 340.49 lakhs in 1971-72 increased to Rs. 1983.10 in 1987-88, this increase could only be due to increase in the prices of MFP (Table 4.5).

Table 4.5 clearly indicate that tribals' share in the exploitation of forest produce has always been marginal. With increasing commercialisation of forest produce on the one hand, and declining forest-cover on the other, it is becoming more and more difficult for the tribals to subsist on the forest produce available to them.

TABLE 4.5: Exploitation of Forest Produce by Different Agencies

Year	Agency of Exploitation	Timber ('000 cu.mt.)	Fuelwood ('000 cu.mt.)	MFP (lakh Rs)
1971-72	Forest Department	251.022 (58.32)	624.647 (36.62)	261.11 (76.68)
	Purchasers	174.099 (40.45)	1006.284 (59.00)	76.43 (22.44)
	Rights and Privilege-holders	5.234 (1.21)	74.385 (4.36)	2.95 (0.86)
	Total	430.364 (100.00)	1705.316 (100.00)	340.49 (100.00)
1975-76	Forest Department	154.589 (52.63)	351.813 (40.28)	262.04 (66.59)
	Purchasers	136.391 (46.43)	502.788 (57.57)	125.68 (31.93)
	Rights and Privilege-holders	2.725 (0.92)	19.680 (2.25)	5.79 (1.48)
	Total	293.705 (100.00)	873.281 (100.00)	393.51 (100.00)
1980-81	Forest Department	168.760 (57.08)	467.342 (40.71)	368.46 (70.21)
	Purchasers	125.867 (42.57)	638.855 (55.64)	146.23 (27.86)
	Rights and Privilege-holders	1.017 (0.34)	41.880 (3.64)	10.09 (1.92)
	Total	295.644 (100.00)	1148.167 (100.00)	524.78 (100.00)
1984-85	Forest Department	158.682 (61.51)	354.249 (41.58)	1217.74 (49.24)
	Forest Labour Cooperative Societies (FLCS)	94.738 (36.72)	396.581 (46.54)	1223.48 (49.47)
	Rights and Privilege-holders	4.541 (1.76)	101.126 (11.86)	31.72 (1.28)
	Total	257.961 (100.00)	851.956 (100.00)	2472.94 (100.00)
1987-88	Forest Department	191.004 (68.15)	246.167 (42.69)	1508.10 (76.05)
	Forest Labour Cooperative Societies (FLCS)	87.992 (31.39)	292.887 (50.79)	445.17 (22.45)
	Rights and Privilege-holders	1.282 (0.46)	37.528 (6.52)	29.82 (1.50)
	Total	280.278 (100.00)	576.582 (100.00)	1983.10 (100.00)

Note. Agencywise exploitation breakup of MFP for 1985-86 is not available.
Source: Forest Statistics of Maharashtra, *op. cit.*

Sample of the Study

To study the economic impact of the Governments forest policy, three types of samples were used for the field-survey. They are as follows:

(1) Randomly selected households living in purposively selected forest villages of Maharashtra;
(2) Forest officials ranging from Range Forest Officer to Conservator of Forests or equivalent, from the Forest Department of Maharashtra, the Forest Development Corporation of Maharashtra, and the Department of Social Forestry in Maharashtra;
(3) A group of 'typical villages' —

   (a) an unauthorised settlement,
   (b) an ex-proprietary forest village,
   (c) a village settled by the Tribal Development Corporation of Maharashtra,
   (d) a Kathiawadi settlement, and
   (e) an irrigation-project settlement created by the Revenue Department.

Methodology used for Sample-Selection

Sample 1

From the eleven forest circles of Maharashtra (Fig. 4.1), four circles, namely, Amravati, Dhule, Thane and South Chandrapur, were purposively selected for the reasons given below.

Amravati

Situated in the Central-East part of Maharashtra, Amravati has a large area of thick forest cover (Fig. 4.2). In the Melghat region, the 'in-forest settlements' (i.e. forest villages) were made for the first time in Maharashtra. It also has large tribal population mainly of the Korkus.

Dhule

Dhule circle, in the north of the State, was bifurcated from Nasik circle in 1983 (Fig. 4.3). It covers all the major forest area in Satpura ranges. There are many ex-proprietary forest villages in this circle with majority of tribal population, belonging mainly to Bhil, Kokna, Pardhi and other tribes.

Thane

It is on the Western side of the State. The forest in this circle, though degraded by now, covers large geographical area and is dominated by politically aware Warlis and Katkari tribes (Fig. 4.4).

South Chandrapur

This circle is situated in South-Eastern Maharashtra. Chandrapur circle was divided into two circles, namely, the North Chandrapur and the South Chandrapur (Fig. 4.5). The rich forest of Allapalli, Bhamragarh and Sironcha is incorporated in the South Chandrapur circle. It has large forest area, the largest in Maharashtra, with teak and other mixed species. Majority of the population is tribal, mainly Gond and Madia.

From these four circles, forest villages were purposively selected. Forest villages are the tribal — settlements made by the British to ensure supply of labour for various forest works. Till 1977, there were 447 forest villages in Maharashtra, with more than one lakh population mainly tribals. These tribals were given a piece of land for cultivation, on lease, and were entitled for some other concessions on forest produce. In return, they had to work for the Forest Department on wages. The tribals residing in these forest villages came in closer contact with the working of the department (than the tribals in other revenue villages)and were affected most, by any change in the forest policy. Thirteen years back, all the forest villages were converted into 'revenue villages' (G.R.No. FLD 3675/7819 — F_6 Dated 24-8-1977).

Thus the villages which were previously 'forest villages', were selected from the four circles mentioned above, for specific reasons. After visiting the circle offices, and looking from the map, the expanse of the circle, location of villages, and taking into

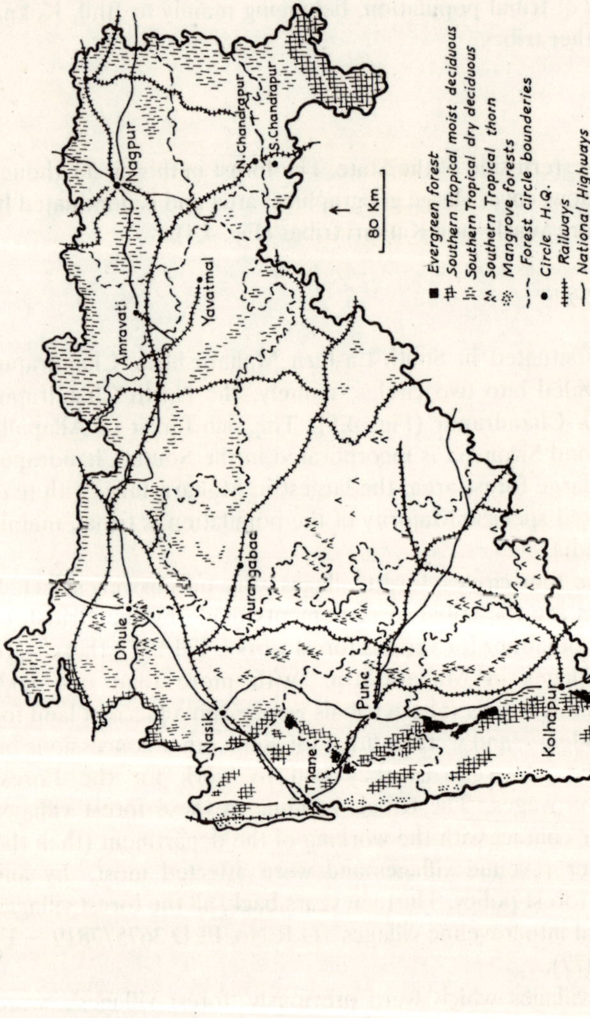

FIG. 4.1: Forest Circles in Maharashtra.

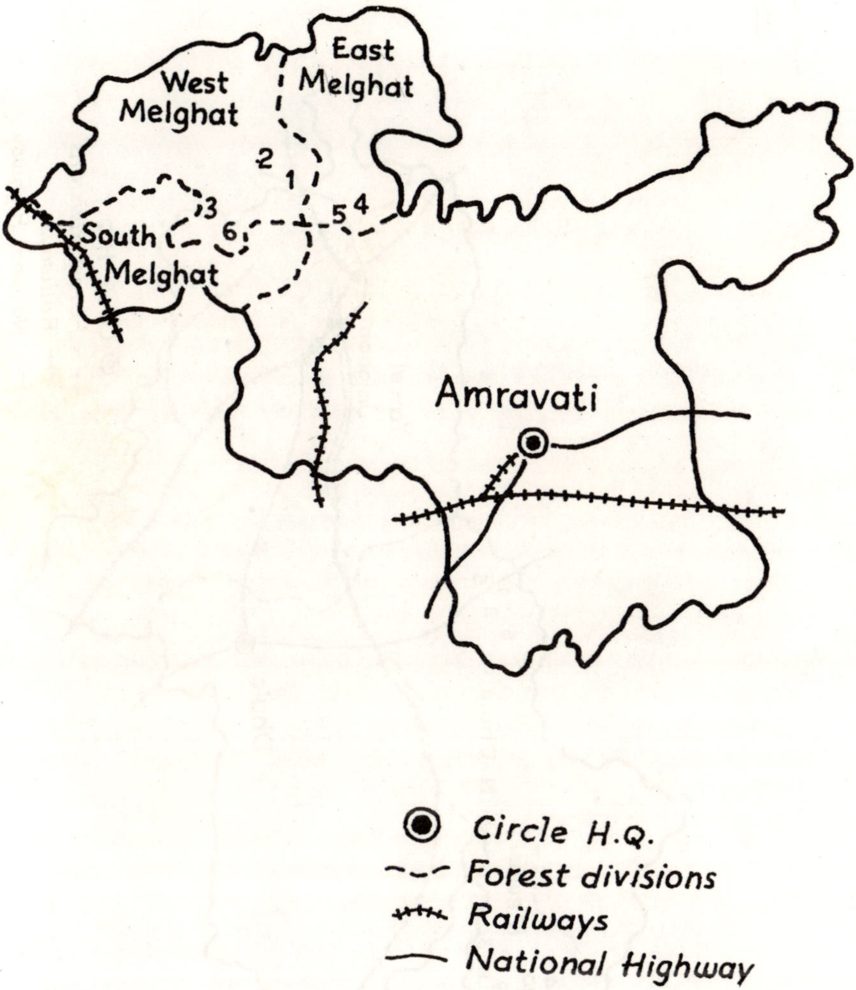

FIG. 4.2: Amravati Circle.

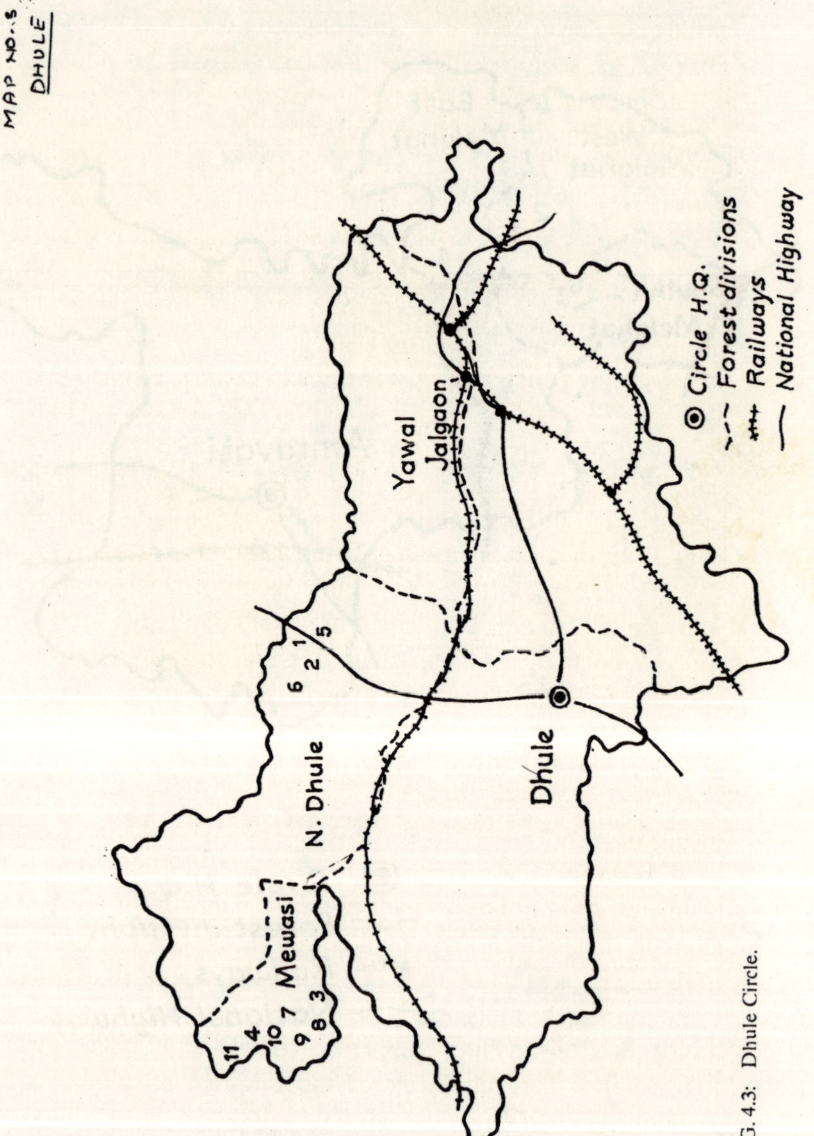

FIG. 4.3: Dhule Circle.

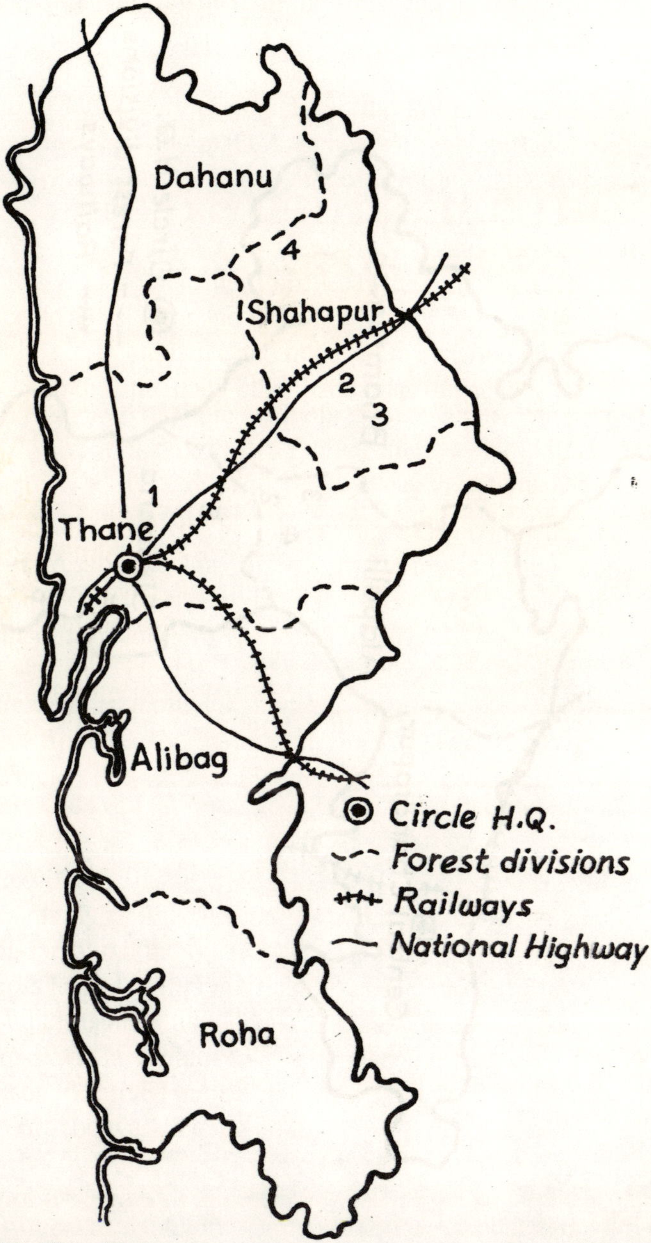

FIG. 4.4: Thane Circle.

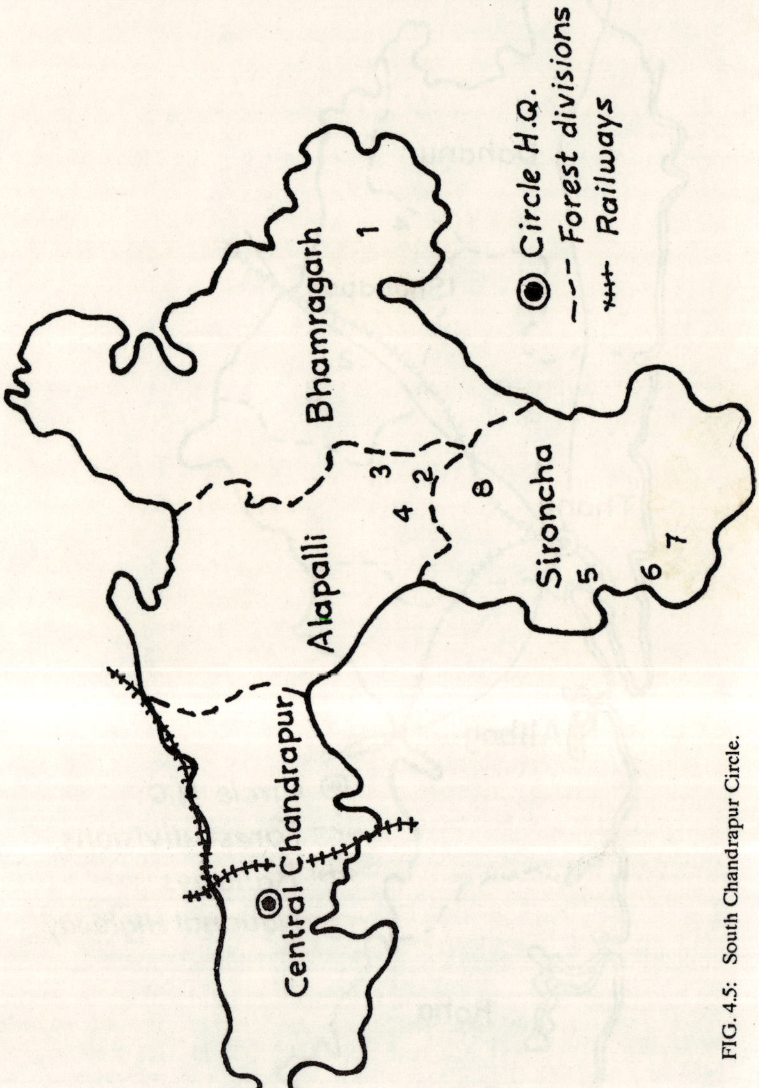

FIG. 4.5: South Chandrapur Circle.

consideration the desirability of visiting particular clusters of villages etc., the number of sample villages was arrived at. The selection of specific villages to be visited was done by getting requisite information from the staff of the Forest Department so as to have representative villages from those circles. The sample thus constitutes of 23 forest villages. Maps showing forests in Maharashtra and sample villages in the selected four circles are attached in the end of the chapter. The question schedule used for interviewing house-holds from the forest villages, is also attached in Appendix I.

Sample 2

It was decided to cover forest officers working in the field as well as on decision-making posts from the three departments working in forests in the State, namely, the Forest Department, the Forest Development Corporation of Maharashtra, and the Department of Social Forestry. In all, 10 officers were interviewed. Two Conservators of Forests, 2 Divisional Forest Officer, 2 Assistant Conservators of Forests and 1 Range Forest Officer — all from the Forest Department. Two Officers from the Forest Development Corporation of Maharashtra, and one officer from the Department of Social Forestry, were also interviewed with the help of pre-tested question schedule.

Major limitation of the sample is that it is not large enough to claim representative character. Yet, it gives some indication of the opinions prevailing among those who are responsible for forest development and come in contact with the people. The respondents were not randomly selected, but on the basis of their general interest in the tribal problems and their willingness to be interviewed.

The question schedule used for this sample is attached in Appendix II.

Sample 3

Though this sample was not in the scheme earlier, after the discussion with the Forest Department staff and at the time of field-survey, certain villages or settlements, which were not forest villages, were found to be peculiar and of profound importance from the point of view of the study. All the five such typical clusters were studied by selecting one from each of the categories, namely, unauthorised settlements, ex-proprietary forest villages, settlements made by the Tribal Development Corporation of Maharashtra, Kathiawadi villages, and irrigation-project settlements created by the Revenue Department. Details about these villages were collected, observations were made and few households were interviewed (Fig. 4.1 to 4.5).

Limitations of the Samples

(1) It was planned to collect the list of villagers from the *sarpanch* of the respective villages for the random selection of households. The list, however, could not be made available and on approaching the villages, all the villagers used to gather around. It was virtually impossible to select them beforehand. From this gathering five persons were randomly selected for the interview.
(2) In Chandrapur, Dhule and Amravati, help of interpreters was taken and therefore questions could not be put directly.
(3) Respondents were reluctant to disclose information regarding encroachments, and their association with social organisations.
(4) The respondents were, at times, unable to give specific information regarding age of their family members, number of cattleheads, land owned, land encroached, annual quantity produced, fuelwood requirement etc.
(5) Majority of the villagers could not differentiate between post-Independence and pre-Independence conditions. Many were not aware of the difference between a forest-village and a revenue village.
(6) There are many agencies affecting the socio-economic conditions of the tribals, such as money-lenders, middlemen, Tribal Development Corporation, Forest Labour Cooperative Societies etc. But the impact of forest policy and the working of the Forest Department alone was considered for the present study as a dominant factor.

Summary

Information regarding geo-climatic features of Maharashtra, the tribal areas and tribal population is given in the beginning of this

Chapter. The details of forest area in the State and its administration have also been given. The rights and privileges of the forest dwellers are enumerated as well.

Discussion of the three samples used for field survey, their selection, and methodology of the study has also been undertaken. Lastly, the limitations of the field survey have been presented.

The results of the field survey are discussed in the next chapter.

NOTES

1. Compiled from —
   - (i) Report of the Committee for Afforestation of Barren, Denuded Lands and Hills of Maharashtra, (1982), Government of Maharashtra.
   - (ii) *Gazetteer of India*, Maharashtra State, Thane, Dhule, Amravati and Chandrapur districts.
2. Report on Development of Tribal Areas, National Committee on the Development of Backward Areas, (1981), Planning Commission, Government of India, New Delhi.
3. 'Maharashtra — Organising the Adiwasis', *Economic and Political Weekly*, Vol.IX Nos. 6,7,8 Feb. 1974. (Annual Number), p. 174-A.
4. *Bombay Forest Manual*, Vol. III, By R.B.D.V. Koppikar, 3rd edition (1950), Government of Bombay. p. 56.
5. *Ibid.*, p. 56
6. *Ibid.*, p. 56
7. *Ibid.*, p. 56
8. Compiled from :
   - (i) The *Bombay Forest Manual*, Vol. III, Third edition, (1950), Government of Bombay Presidency.
   - (ii) Concessions and Plans Regarding Forests in Maharashtra, Tribal Research and Training Institute, Pune.
   - (iii) Working Scheme, For the Gatta, Tadgaon and Bhamragarh Ranges of Bhamgarh Division, Chandrapur Circle, Maharashtra State, for the years 1970-71 to 1979-80. Vol. I (Text), Government of Maharashtra.
   - (iv) Working Plan for Reserved Forests of Thane district, Thane Circle. For the years 1969-70 to 1988-89, Vol. II, Government of Maharashtra.
   - (v) Working Plan for the forests of North-Dhulia Division, Vol. I, Part I, For the years 1970-71 — 1989-90, Government of Maharashtra.

5

FINDINGS OF THE FIELD-WORK

This chapter is devoted to the discussion on the findings of the field survey. The chapter is divided into three sections. The first section deals with the sample of households, randomly selected from the forest villages. Economic and demographic features of the households, civic amenities available in the selected forest villages, are discussed in details. The second section constitutes the findings from the sample of forest officers. The results are based on the interviews often forest officials. Section three deals with the sample of 'typical villages'. These villages have some peculiar characteristics, important from the point of view of the study, and, hence, were included in the study. Inferences drawn from the study of the three samples are given at the end of respective section.

SECTION I

SAMPLE OF HOUSEHOLDS FROM FOREST VILLAGES

The process of the selection of forest villages and the households therefrom, has already been discussed in the previous chapter. The observations made from the interviews of the households are given below:

Amenities in the Sample Villages

For the sample villages, nine types of civic amenities were specified to judge their status. These were:

(1) fair-price shop,
(2) primary health centre,
(3) electricity,
(4) potable water,

(5) primary school,
(6) independent *gram-panchayat*,
(7) bank,
(8) post office, and
(9) tar-road.

It was observed that out of these nine amenities, not more than four were available in any of the 23 sample villages. Other significant observations are:

(1) None of the villages had either a post office or a bank.
(2) Solar street-lights were provided only in two villages, and no other village had electricity.
(3) Except in three villages which had tar road, all other villages had *kuccha* mud roads which were in very bad conditions, thus isolating the village in rainy seasons.
(4) Only two villages had fair-price shops and the same two villages (only) had the primary health centres.
(5) Fifteen villages had the facility of potable water either through wells or hand pumps. But these villages are mainly from South Chandrapur and Amravati Circles. This facility was very poor in Dhule circle where the sample villages received water only on alternate days by tankers (on account of draught).
(6) Only five villages had primary schools but most of them were reported to be without a teacher at the time of the survey.
(7) Except for three villages which had group gram-panchayats, all other twenty villages had independent *gram-panchayats*.

This list of sample villages with the number of civic amenities available (out of the nine amenities specified above), is given in Table 5.1.

If the distance of these villages from all-weather roads and nearest market is considered, the plight of the sample villages becomes clear to some extent. While 10 villages were 5 to 10 km away from nearest market, one village was as much as 30 km away from a market. Eleven villages were 10 to 25 km from the market, and only one village was within 5 km of the nearest market.

In case of distances from all-weather road, 9 villages were between 10 to 20 km, 7 villages were within 5 km, and 6 villages were as far as 10 to 20 km. The farthest single village was between 25–30 km away from an all-weather road. Table 5.2 denoting the distance from nearest market and all-weather road for the sample villages is given.

TABLE 5.1: Number of Civic Amenities Available in Sample Villages

Sl. No.	Name of Circle	Name of village	Number of amenities available
1.	Amravati	Raipur	2
2.	Amravati	Hatru	2
3.	Amravati	Kund	2
4.	Amravati	Keli	1
5.	Amravati	Khongada	2
6.	Dhule	Khairkuti	1
7.	Dhule	Bijrigaon	4
8.	Dhule	Gadhaddeo	2
9.	Dhule	Decaty	1
10.	Dhule	Ekdhad	2
11.	Dhule	Pimparpada	1
12.	Dhule	Ratanpada	3
13.	Thane	Ambarbhui	2
14.	Thane	Tilmal	3
15.	Thane	Satronda	4
16.	Thane	Airc	Nil
17.	South Chandrapur	Yenkabanda	2
18.	South Chandrapur	Rompally	4
19.	South Chandrapur	Maraigudam	Nil
20.	South Chandrapur	Talwada	3
21.	South Chandrapur	Zamela	3
22.	South Chandrapur	Yelchil	3
23.	South Chandrapur	Yergudtola	2

Demographic Features

(a) *Family Size*

It was observed that more than half (58.26%) of the sample households had family members between 6 to 10 followed by 26.96 per cent of the sample households having less than five family members. 14.78 per cent of the sample households had 11 or more family members.

The households from Thane circle had as high as 70 per cent of the households having family members numbering between 6 and 10.

South Chandrapur circle had very few (i.e. only 08.57%) households with members more than 10 (Table 5.3).

TABLE 5.2: Distance of Sample Villages from All-weather Roads and Markets

Sl. No	Distance (in km)	Number of sample villages	
		From all-weather roads	From the nearest market
1.	0-5	7	1
2.	5-10	9	10
3.	10-15	3	4
4.	15-20	3	5
5.	20-25	Nil	2
6.	25-30	1	1
	Total	23	23

TABLE 5.3: Family Size of Sample Households

Sl. No.	Name of Circle	Number of households with family members			Total
		between 1-5	between 6-10	above 10	
1.	Amravati	6 (24.00)	14 (56.00)	5 (20.00)	25 (100.00)
2.	Dhule	11 (31.44)	19 (54.28)	5 (14.28)	35 (100.00)
3.	Thane	2 (10.00)	14 (70.00)	4 (20.00)	20 (100.00)
4.	South Chandrapur	12 (34.28)	20 (57.15)	3 (08.57)	35 (100.00)
	Total	31 (26.96)	67 (58.26)	17 (14.78)	115 (100.00)

Note: Figures in brackets indicate percentage to the horizontal totals.

(b) *Sex Ratio*

The sex ratio for the sample was 1000 : 914 which is close to the sex ratio of Maharashtra namely, 1000 : 937 and that of India which is 1000 : 933.

It was low for Thane circle (1000 : 822) and slightly high for Amravati circle (1000 : 967). The sex ratio was near the average for Dhule (1000 : 953) and South Chandrapur (1000 : 902) circles (Table 5.4).

TABLE 5.4: Sex Ratio in the Four Circles

Sl. No.	Name of Circle	Number of members			Sex Ratio
		Male	Female	Total	
1.	Amravati	92	89	181	1000 : 967
2.	Dhule	129	123	252	1000 : 953
3.	Thane	90	74	164	1000 : 822
4.	South Chandrapur	122	110	232	1000 : 902
	Total	433	396	829	1000 : 914

(c) *Educational Status*

From the sample 84.44 per cent of the total family members were found to be illiterate, illiteracy being highest in South Chandrapur circle (94.40%) and lowest in Dhule circle (74.40%). Out of the sample, 12.67 per cent of the family members had primary education and 02.89 per cent had received secondary education. Not a single person had college education.

In Amravati and South Chandrapur circles no one had received even secondary education. In Thane circle 3 persons (1.83%), and 21 persons (8.33%) from Dhule circle had education up to the secondary standard (Table 5.5).

(d) *Housing Conditions*

It was a miserable sight in the villages to find tribals in their age-old thatched roofs and small houses even after years of Independence.

(i) *Area*

Out of the sample of 115 households 87 (i.e. 75.65%) households had only 200 to 400 sq. ft. area for their huts. Only 28 (i.e. 24.35%) of the huts were more than 400 sq.ft. in area.

In the South Chandrapur circle, all the 35 households had less than 400 sq.ft. of housing area. It was only in Dhule circle that 16 out of 35 households (i.e. 45.71%) had housing area more than 400 sq.ft. (Table 5.6)

TABLE 5.5: Educational Status of Family Members of Sample Households

Sl. No.	Name of Circle	No. of family members			Total
		Illiterate	Primary	Secondary	
1.	Amravati	159 (87.85)	22 (12.15)	Nil	181 (100.00)
2.	Dhule	190 (75.40)	41 (16.27)	21 (08.33)	252 (100.00)
3.	Thane	132 (80.49)	29 (17.68)	3 (01.83)	164 (100.00)
4.	South Chandrapur	219 (94.40)	13 (05.60)	Nil	232 (100.00)
	Total	700 (84.44)	105 (12.67)	24 (02.89)	829 (100.00)

Note: Figures in brackets indicate percentage to horizontal total.

TABLE 5.6: Number of Houses According to Area

Sl. No.	Name of Circle	Area of Houses		Total
		200-400 sq.ft.	above 400 sq.ft.	
1.	Amravati	20 (80.00)	05 (20.00)	25 (100.00)
2.	Dhule	19 (54.29)	16 (45.71)	35 (100.00)
3.	Thane	13 (65.00)	7 (35.00)	20 (100.00)
4.	South Chandrapur	35 (100.00)	Nil	35 (100.00)
	Total	87 (75.65)	28 (24.35)	115 (100.00)

Note: Figures in bracket indicate percentage to horizontal total.

(ii) *Roof Types*

Local tiles, Manglore tiles, and Grass and reed — these three types of materials were used for roofs of the tribal huts. In the sample, it was found that 65.22 per cent of the households had local, home-made tiles, 22.60 per cent had Manglore tiles and 12.18 per cent of the total households had roofs thatched with grass and reed.

The peculiar observation is, that in Thane circle there were no local tiles, in Amravati circle no household used Manglore tiles and in Dhule circle no hut had thatched roof.

South Chandrapur circle had as many as 8 households, out of 35, (i.e. 22.86%) which had huts with roofs made of grass and reed. 22 out of 25 households in Amravati circle used local tiles for roofs of their huts (Table 5.7).

TABLE 5.7: Types of Roofs

Sl. No.	Name of Circle	Local tiles	Mangalore tiles	Grass and reed	Total
1.	Amravati	22	Nil	3	25
		(88.00)		(12.00)	(100.00)
2.	Dhule	33	2	Nil	35
		(94.29)	(05.71)		(100.00)
3.	Thane	Nil	17	3	20
			(85.00)	(15.00)	(100.00)
4.	South Chandrapur	20	7	8	35
		(57.14)	(20.00)	(22.86)	(100.00)
	Total	75	26	14	115
		(65.22)	(22.60)	(12.18)	(100.00)

Note: Figures in brackets indicate percentage to horizontal total.

Economic Features

(a) *Occupation*

Occupations were classified as cultivator, general labour, agricultural labour, and forest labour. The groups were not found to be mutually exclusive. It was observed that out of 115 respondents only 5 (i.e. 4.35%) were just cultivators. All others worked as labourers in one form of the other. As many as 97 (i.e. 84.35%) were cultivators as well as general labourers and forest labourers. Four (i.e. 3.48%)

respondents each belonged to the three combinations of — 'agricultural labour and forest labour', 'general, agricultural and forest labour' and 'cultivator, general, agricultural and forest labour'. Only 1 (i.e. 0.86%) respondent belonged to the category of 'cultivator and forest labourers'.

In all there were eight landless respondents three each from Thane (15.00%) and Dhule (8.5%) circles and one each from Amravati (4.00%) and South Chandrapur (2.86%) circles.

All the five respondents who depended on cultivation alone for their livelihood belonged to Dhule circle (14.28%).

From the sample eight (6.96%) respondents worked as agricultural labourers, out of which seven were from Dhule circle and one was from Amravati circle. There were no agricultural labourers either in Thane or in South Chandrapur circles (Table 5.8).

TABLE 5.8: Occupational Structure

Occupation Category	Number of respondents				Total
	Amravati	Dhule	Thane	South Chandrapur	
1	–	5 (14.28)	–	–	5 (04.35)
1+4	–	1 (2.86)	–	–	1 (00.86)
2+4	–	–	3 (15.00)	1 (2.86)	4 (03.48)
1+2+4	24 (96.00)	22 (62.86)	17 (85.00)	34 (97.14)	97 (84.35)
2+3+4	1 (04.00)	3 (08.58)	–	–	4 (03.48)
1+2+3+4	–	4 (11.42)	–	–	4 (03.48)
Total	25 (100.00)	35 (100.00)	20 (100.00)	35 (100.00)	115 (100.00)

Notes :(1) The codes used are —
1. Cultivator 2. General labour
3. Agricultural labour 4. Forest labour
(2) Figures in brackets indicate percentage to vertical totals.

(b) *Land Ownership*

As already mentioned earlier, most of the sample households owned land because Forest Department had given some land to each households at the time of their settlement in the forest villages. From the sample 68.69 per cent of the households possessed land up to 2 ha, 17.39 per cent between 2 to 4 ha, while only 06.96 per cent had more than 4 ha of land in their name. The percentage of landless households was 6.96 only.

In Thane and South Chandrapur circles not a single person possessed land more than 4 hect. In Dhule circle 6 households (i.e 17.14%) and 2 households (i.e. 8.00%) from Amravati circle possessed more than 4 hect. of land (Table 5.9).

TABLE 5.9: Land Ownership Pattern

Sl. No.	Name of Circle	No. of households owning land				Total
		Up to 2 ha	between 2-4 ha	above 4 ha	Landless	
1.	Amravati	16 (64.00)	6 (24.00)	2 (8.00)	1 (4.00)	25 (100.00)
2.	Dhule	20 (57.14)	6 (17.14)	6 (17.14)	3 (8.58)	35 (100.00)
3.	Thane	14 (70.00)	3 (15.00)	–	3 (15.00)	20 (100.00)
4.	South Chandrapur	29 (82.86)	5 (14.28)	–	1 (2.86)	35 (100.00)
	Total	79 (68.69)	20 (17.39)	8 (6.96)	8 (6.96)	115 (100.00)

Note: Figures in brackets indicate percentages to horizontal total

(c) *Encroachments*

As is clear from the above discussion, most of the households had land which was less than two hectares. The insufficiency of the land owned and its low productivity on the one hand and large extent of immigration along with decreasing availability of forest produce on the other, has induced tribals to encroach on the forest land for extending their cultivation.

If the number and extent of encroachments is considered in all the four circles, the results of the survey indicate that out of 115 households, 44 households (i.e 38.27%) had not encroached. 51 households (i.e. 44.34%) had encroached on more than two ha and only 20 households (i.e. 17.39%) had encroached more than two ha but less than three ha of forest land. Not a single individual had encroached more than three ha of land.

It was only in Dhule Circle that as high as 28.57 per cent of the households had encroached more than two ha of forest land. The percentage of households under this category in Amravati circle was 8.00, in Thane Circle 15.00, and 14.28 in South Chandrapur Circle, which is comparatively low (Table 5.10).

TABLE 5.10: Number of Households with or without Encroachment

Sl. No.	Name of Circle	Number of Households with encroached area		No. of households without encroachment	Total
		Up to 2 ha	Between 2-3 ha		
1.	Amravati	14 (56.00)	2 (8.00)	9 (36.00)	25 (100.00)
2.	Dhule	5 (14.28)	10 (28.57)	20 (57.15)	35 (100.00)
3.	Thane	7 (35.00)	3 (15.00)	10 (50.00)	20 (100.00)
4.	South Chandrapur	25 (71.44)	5 (14.28)	5 (14.28)	35 (100.00)
	Total	51 (44.34)	20 (17.39)	44 (38.27)	115 (100.00)

Note : Figures in brackets indicate percentages to horizontal totals.

(d) *Livestock Ownership*

It was a general observation that majority of the tribal households owned cattle. Out of the sample of 115 households only 6 households (1 from Amravati; 2 from Dhule and 3 from South Chandrapur circles) were without any cattle (Table 5.11).

(e) *Dependence on Forests*

The three categories regarding the question of depedence on forests were — no dependence, marginal dependence and substantial dependence. Wherever there was no real tree cover around the villages and the villages could not get any forest produce, they were included in the first category of 'no dependence'. 'Marginal dependence' meant, where the villagers could get fuelwood, fodder and some fruits — roots etc., but only for self-consumption. The category of 'substantial dependence' was defined as, where the villagers could manage some livelihood by selling forest produce in addition to self-consumption. It also provided them with some remunerative employment (Table 5.12).

TABLE 5.11: Ownership of Livestock

Sl. No.	Name of Circle	Average number of cattlehead per household			
		Bullock	Cows	Goats	Buffaloes
1.	Amravati	3.04	2.24	4.56	0.24
2.	Dhule	2.80	2.51	4.23	0.14
3.	Thane	2.60	2.30	3.75	Nil
4.	South Chandrapur	3.34	2.77	4.40	0.40
	Total	2.98	2.49	.4.29	0.21

It was observed that out of the total of 115 households, 71.30 per cent i.e. 82 households depended on forest marginally, 20.00 per cent i.e. 23 households did not get anything from the forest, while 8.70 per cent i.e. 10 households only, depended 'substantially' on forests.

It is very important to note that in half of the sample, i.e. in the Thane and Dhule circles, not a single respondent depended on forests substantially. It is only from Amravati and South Chandrapur circles that five persons each (i.e. 20.00 per cent and 14.29 per cent respectively) could earn some livelihood from the forests (Table 5.12).

(f) *Awareness of Privileges and Concession*

It was observed that the tribal respondents were mostly ignorant of their rights, privileges and concessions available on the use of forest produce. They were using the forest produce traditionally but were not aware of the legal sanctions regarding quantity of fuelwood which can be collected without fees, grazing fees to be paid on each cattle-unit, or the fruits which can be collected without paying any price etc. As many as 97.40 per cent of the respondents showed lack of knowledge regarding the details of such sanctions.

This ignorance was total amongst the respondents from South

Chandrapur and Amravati circles. Only 10 per cent of the respondents in Thane and 2.86 per cent in Dhule circles showed some awareness in these regards (Table 5.13).

TABLE 5.12: Dependence on Forests

Sl. No.	Name of Circle	Number of households depending on forests			Total households
		No dependence	Marginal dependence	Substantial dependence	
1.	Amravati	2 (8.00)	18 (72.00)	5 (20.00)	25 (100.00)
2.	Dhule	6. (17.14)	29 (82.86)	Nil	35 (100.00)
3.	Thane	15 (75.00)	5 (25.00)	Nil	20 (100.00)
4.	South Chandrapur	Nil	30 (85.71)	5 (14.29)	35 (100.00)
	Total	23 (20.00)	82 (71.30)	10 (8.70)	115 (100.00)

Note: Figures in brackets indicate percentages to horizontal total.

TABLE 5.13: Awareness of Privileges

Sl. No.	Name of Circle	Number of Respondents		
		Aware	Unaware	Total
1.	Amravati	Nil	25 (100.00)	25 (100.00)
2.	Dhule	1 (2.86)	34 (97.14)	35 (100.00)
3.	Thane	2 (10.00)	18 (90.00)	20 (100.00)
4.	South Chandrapur	Nil	35 (100.00)	35 (100.00)
	Total	3 (2.60)	112 (97.40)	115 (100.00)

Note: Figures in brackets indicate percentages to horizontal total.

(g) *Problems in Using Forest Produce*

As mentioned earlier the tribal households were using forest produce since time immemorial. But they are facing some problems in forest produce now-a-days. It was found that mainly four problems were confronted in this regard. These four problems are: (1) lack of information regarding the rights and concessions over forest produce; (2) exploitation by the forest department; (3) distance; and (4) non-availability of forest produce. It was found that in all, not more than six types of combinations of these problems were faced by the sample households. The details are as given below.

Thirty-five households (i.e. 30.43%) faced the only problems of lack of information. These consisted of 15 households from Amravati circle (i.e. 60.00%) and 20 from South Chandrapur circle (i.e. 57.15%).

Fifteen households (i.e. 13.04%) had both the lack of information and exploitation problems. These consisted of five households from Amravati circle (i.e. 20.00%) and 10 households from South Chandrapur circle (i.e. 28.57%).

Only five households (i.e. 4.35%) suffered from the lack of information and distance problem. All these belonged to Amravati circle (i.e. 20.00%).

Five households (i.e 4.35%), all belongings to South Chandrapur circle, found lack of information, exploitation and distance together, as their problems.

Three households of the sample (i.e. 2.60%) suffered from exploitation, distance and non-availability of forest produce. These consisted of two households from Thane circle (i.e. 10.00%) and one household from Dhule circle (i.e. 2.85%).

A very large number of households i.e. 52 (45.22%) suffered from all the four above-mentioned problems. These consisted of 18 households from Thane circle (i.e. 90.00%) and 34 households from Dhule circle (i.e. 97.15%) (Table 5.14).

(h) *Exploitation by Forest Department Employees*

It is clear from the previous discussion that 'exploitation' was a major problem faced by the tribal respondents. Usually the nature of exploitation as reported was —

(i) less payment of wages than sanctioned;
(ii) lower wages or, at times, no wages to female labourers;
(iii) instead of insisting on grazing licences, the staff of forest department asked for bribes from the cattle owners;
(iv) as discussed earlier, the households of forest villages are entitled to fixed quota of housing material, fuelwood, fodder and other minor forest produce, either free of cost or at nominal price. Misusing the ignorance of the tribals about these concessions and privileges, the forest department employees extracted bribes, in cash or kind, from the tribals if they were 'caught' while taking out the forest produce.

TABLE 5.14: Problems in using Forest Produce

Problem Type	Number of Respondents				Total
	Amravati	Dhule	Thane	South Chandrapur	
A	15 (60.00)	–	–	20 (57.15)	35 (30.43)
A+B	5 (20.00)	–	–	10 (28.57)	15 (13.04)
A+C	5 (20.00)	–	–	–	5 (04.35)
A+B+C	–	–	–	5 (14.28)	5 (04.35)
B+C+D	–	1 (2.85)	2 (10.00)	–	3 (2.60)
A+B+C+D	–	34 (97.15)	18 (90.00)	–	52 (45.22)
Total	25 (100.00)	35 (100.00)	20 (100.00)	35 (100.00)	115 (100.00)

Note: (1) The codes used : (A) Lack of information regarding rights and concessions, (B) exploitation by the forest department, (C) distance, (D) non-availability of forest produce.

(2) Figure in brackets indicate percentage to vertical totals.

Regarding the first three types of exploitation enumerated above, first hand information could not be collected. But, when asked about their indulgence in bribing the forest employees, more than half, i.e. 65.21 per cent of the respondents answered in affirmative. The percentage was very high in Dhule and Thane circles, while it was low in Amravati and South Chandrapur circles (Table 5.15).

TABLE 5.15: Households involved in Bribery

Sl. No.	Name of Circle	Number of households involved		Total
		Yes	No	
1.	Amravati	5 (20.00)	20 (80.00)	25 (100.00)
2.	Dhule	35 (100.00)	Nil	35 (100.00)
3.	Thane	20 (100.00)	Nil	20 (100.00)
4.	South Chandrapur	15 (42.85)	20 (57.15)	35 (100.00)
	Total	75 (65.21)	40 (34.79)	115 (100.00)

(i) *Wages in Forestry and Agriculture*

Wages paid by the Forest Department are fixed and known to all, and there is no vagueness about it. When asked about the difference in these wages and those of agricultural labourers (which changed according to time and place), 53.40 per cent of the total respondents accepted that forest wages were better than agricultural wages. But the picture differed from circle to circle.

In Thane circle 100.00 per cent of households reported agriculture wages to be higher than forest wages. The percentage of respondents saying so was considerably high in Dhule Circle as well (82.86%). But in the South Chandrapur and Amravati circles 100.00 per cent and 80.00 per cent of the respondents, respectively, found the wages paid by the forest department to be better than agricultural wages (Table 5.16).

(j) *Forest Villages and Revenue Villages*

As has been explained earlier, residents of forest villages enjoyed certain special privileges regarding the use of forest produce. But, for several reasons, the villages were converted into revenue villages. The sample households were asked about their preference about forest and revenue villages. Out of the sample of the 115 households, 771 (i.e. 61.75 per cent were in favour of forest villages and 29 (i.e.

25.21%) were in favour of revenue villages, while the remaining 15 (i.e. 13.04%) could not differentiate between the two (all of them belonged to Amravati Circle).

TABLE 5.16: Comparative Opinion about Agriculture and Forest Wages

Sl. No.	Name of Circle	Number of respondents in favour of		Total
		Forest Department wages	Agriculture wages	
1.	Amravati	20 (80.00)	5 (20.00)	25 (100.00)
2.	Dhule	6 (17.14)	29 (82.86)	35 (100.00)
3.	Thane	Nil	20 (100.00)	20 (100.00)
4.	South Chandrapur	35 (100.00)	Nil	35 (100.00)
	Total	61 (53.04)	54 (46.96)	115 (100.00)

Note: Figures in brackets indicate percentages to horizontal totals.

While 88.57 per cent from South Chandrapur circle preferred forest villages, not a single respondent was in favour of forest village in Thane circle (Table 5.17).

TABLE 5.17: Forest Villages and Revenue Villages

Sl. No.	Name of Circle	No. of households in favour of		No answer	Total
		Forest village	Revenue village		
1.	Amravati	10 (40.00)	Nil	15 (60.00)	25 (100.00)
2.	Dhule	30 (85.71)	5 (14.29)	Nil	35 (100.00)
3.	Thane	Nil	20 (100.00)	Nil	20 (100.00)
4.	South Chandrapur	31 (88.57)	4 (11.43)	Nil	35 (100.00)
	Total	71 (61.75)	29 (25.21)	15 (13.04)	115 (100.00)

Note: Figures in brackets indicate percentage to horizontal totals.

(k) *Expectations from the Government*

The final question referred to the expectations of the respondents from the government. Only three items dominated the list. They were — regular employment, irrigation facilities and potable water supply. Out of all possible combinations, only four combinations were found in the sample. Out of 115 respondents 15 (i.e. 13.04%) expected only regular employment from the government while nine (i.e. 7.84%) expected only irrigation facilities to be supplied. Four respondents (i.e. 3.48%) expected both employment and irrigation facilities and as many as 72 respondents (i.e. 62.60%) wanted all the three i.e. employment, irrigation and potable water facilities to be provided by the government.

All the respondents from Thane Circle, and 32 (i.e. 91.43%) respondents from Dhule circle expected the government to provide all the three facilities enumerated above. As many as 15 respondents (i.e. 42.86%) from South Chandrapur circle could not specify their expectations. 40.00 per cent of the respondents in Amravati circle expected that the government should provide regular employment, irrigation and potable water facilities (Table 5.18).

TABLE 5.18: Expectations from Government

Category	Number of respondents from the four circle				Total
	Amravati	Dhule	Thane	South Chandrapur	
1	5 (20.00)	Nil	Nil	10 (28.57)	15 (13.04)
2	9 (36.00)	Nil	Nil	Nil	9 (07.84)
1+2	1 (04.00)	3 (08.57)	Nil	Nil	4 (03.48)
1+2+3	10 (40.00)	32 (91.43)	20 (100.00)	10 (28.57)	72 (62.60)
No Expectation	Nil	Nil	Nil	15 (42.86)	15 (13.04)
Total	25 (100.00)	35 (100.00)	20 (100.00)	35 (100.00)	115 (100.00)

Notes: (1) Categories : 1. Regular Employment
2. Irrigation
3. Potable water

(2) Figures in brackets indicate percentages to vertical totals.

All the questions which appear in the question schedule are not discussed above for several reasons. For example, there was a set of questions whether, the respondent was ever an owner of land and if he lost it, then how and to whom was it lost. However, there was not a single respondent falling in this category. All the landless respondents never owned land.

Quantitative details about the volume and price of the fruits, medicines, fuelwood, grass and grazing, MFP etc. used, could not be ascertained by the tribals.

The sense of master-servant relationship between forest personnel and tribals was so prevalent that the respondents could not conceive of any other relationship.

The respondents could not specify the difference in the employment pattern and details at the time when their village was a forest village from that when their village was declared a revenue village.

The tribals engaged in illegal felling and sale of forest produce, did not know for whom they were really working. They were also reluctant to disclose other details in this regard.

The tribal respondents were not even aware of various social forestry schemes and projects formulated for their benefit. They were concerned with the problems of declining forest cover, decreasing availability of forest produce and insufficient land for cultivation only.

The respondents also refused to disclose their association with any tribal organisation. There was no question of their being aware of any forest legislation.

Inferences

From the study of the demographic and economic features of the sample villages, the following inferences can be arrived at:

(1) Usually tribals have large families. The reasons for comparatively smaller families in South Chandrapur circle may be the popularity of family planning methods.
(2) The sex ratio in Thane circle (1000 : 822) with higher number of male members in the sample households is due to male immigration to the villages in thecircle.
(3) The high percentage of illiteracy amongst tribals in all the four circles is due to lack of educational facilities and their

indifference towards education. But more important is the fact that poverty compels the parents to engage their children in the work such as collection of firewood, other minor forest produce etc.

(4) Alongwith the custom, dwindling availability of housing material is an important reasons for very small huts.

(5) Majority (i.e. 85.00%) of the huts in Thane Circle had Manglore tiles for the roofs. The obvious reasons for this could be the exposure and the nearness of this area to developed areas like Bombay.

(6) Most of the tribals in the working age-group work as forest labour and general labour. Though majority of them own land given by the forest department, due to the inadequacy of land holdings, primitive methods of cultivation and low productivity, they are forced to search for other gainful employment.

(7) Substantial proportion of encroachments found in all the four circles could be because of insufficiency of land owned, and its low productivity. Another plausible reason could be the fact that the land provided by the forest department was further distributed over the next generations resulting in fragmentation.

(8) Cattle is considered as wealth by tribals, specially in the absence of land and other property. This is the reason for the large number of cattle owned by tribal households.

(9) For generations, tribal communities have depended on forests for their necessities. The decreasing availability of the forest produce has also reduced this benefit, adversely affecting the tribal economy. Fruits, fuelwood, fodder, all of which was available at short distance from the tribal villages, has depleted to a great extent. Now the tribal people have to spend much of their time in the collection of forest produce to fulfill their routine necessities.

(10) Though the government have continued a few privileges and have sanctioned some concessions for the tribal communities, very few tribals are aware of them. Very often the tribals collect and make use of the forest produce considering themselves to be encroachers or thieves, unaware of their own rights and privileges.

(11) Another important problem in using the forest produce is its

non-availability. In the areas like Thane and Dhule, where forests are very much denuded, the privileges and concessions are only formal, and not substantive.

(12) Exploitation of the tribals by the forest department employees was reported to be low in South Chandrapur circle due to the fear of the naxalites. All the tribal respondents accepted that before the naxalite movement spread over the circle, they had to bribe the forest employees.

(13) Forest wages are uniform throughout the State, but agricultural wages differ from circle to circle. The reason for higher agricultural wages in Thane and Dhule circles was obviously the developed agriculture and depleted forests in these areas. High availability of forestry work in South Chandrapur and Amravati circles and lesser number of landless labourers in these circles may be the reasons for lower agricultural wages in these areas.

(14) High percentage of respondents preferring revenue village to forest village in Thane circle is clearly due to awareness regarding the benefits available in revenue villages e.g. land ownership, various tribal-welfare schemes etc. In Amravati circle 15 respondents did not know the difference between the two, mainly because they were not covered under any tribal welfare scheme run by the Revenue Department. Another important reason for this lack of awareness was that, the tribals of many such forest villages had not received the documents regarding registration of their agricultural land issued by the Revenue Department.

(15) The facilities for potable water and irrigation were very poor in Dhule and Thane Circles, worsened by the then severe draught.

(16) Majority of the respondents from all the four circles were under employed and expected that the government should provide them regular employment.

SECTION II

SAMPLE OF FOREST OFFICIALS

It was thought desirable to selects some senior officers from the Forest Department who have lot of experience in the working field as well as in the implementation of the policies of the government. In all, 10 forest officials were interviewed. Seven were from the Forest Department, two were from the Forest Development Corporation of Maharashtra (FDCM) and one from the Department of Social Forestry. From the Forest Department (out of 7), two were Conservator of Forests (CF), two were Divisional Forest Officers (DFO), two were Assistant Conservator of Forests (ACF) and one was a Range Forest Officer (RFO). All the seven respondents were from the four circles selected for sample one, namely, Amravati, Dhule, Thane and Sough Chandrapur. The respondents from Forest Development Corporation of Maharashtra and Social Forestry were from Nagpur offices.

It was generally found that while the officials on higher ranks answered questions based on policy matters with greater ease, the others who work mainly in field, found it easier to answer questions related to day-to-day working. Important observations based on the interview of these 10 forest officials are made here.

Tribal Needs

The first question in the question schedule asked whether necessary consideration was given to the needs of the tribals, in governments forest policy.

All the 10 respondents felt that adequate considerations were given in the policy, two respondents, however, added that the considerations were more than necessary, and hence these considerations are responsible for the present denuded state of the forests.

In general all the respondents felt that tribals are getting all that is necessary from forests and do not face any economic problem.

On the· question of curtailment of rights by the British, seven respondents opined that the curtailment of tribals' rights by the British and later on by the Indian Government was only for the better management of forests. The remaining three stressed that the rights of the tribals were not curtailed at all, but were only regulated.

Majority of the respondents felt that the rights of the tribals in forests, were not adversely affected due to the British policy.

Asked if the Government of free India has increased its control on forests, seven respondents answered in affirmative. But they added that such restrictions are necessary due to the increased population pressure and other reasons. Three respondents maintained that the government was not increasing its control on forests. In general majority of the respondents were not against the increasing control of the government.

The question regarding the settlements made by the British officers, six respondents felt that the tribal claims were given justice. Four respondents, however, felt that in many cases tribals' claims were neglected by the British Settlement Officers.

Community Ownership

Not a single respondent was aware of the fact that tribals originally had community ownership of forests. When asked whether the tribals were settled individually or in communities, all accepted that individual settlements were made by the British.

Restrictions in Reserved Forests

'How do the restrictions in Reserved forests affect tribals?' In answer to this question, all the 10 respondents felt that the restrictions in reserved forests do not affect the tribals adversely. They felt that concession given to tribals in such forests are more than necessary to fulfill the tribals' needs.

To the question on as to what the tribals should do once the area is declared as reserved forests, all the 10 respondents answered that the tribals can continue to live and cultivate in that area. None felt that such a reservation would affect tribals in anyway despite the several restrictions that follow, once a forest area is declared as reserved.

Conversion of Tribals into Cheap Labourers

The respondents were asked to express their opinion on the allegation that through the forest policy, the tribals are being ejected from the forest area and therefore are forced to become cheap labourers. One respondent agreed that because of the present policy, the tribals are becoming cheap labourers for the urban employers. The respondent was of the rank of Divisional Forest Officer. All the remaining nine respondents disagreed with this view..

Adverse effects of Concessions

All the 10 respondents felt that the concessions given to the tribals have adverse effects on the forests. They added that over-exploitation of firewood, illicit felling, illegal sale of minor forest produce, encroachments made for cultivation, over-grazing by the cattle, girdling, burning of trees, burning of grass for getting better tendu leaves etc. hamper the development of forests to a large extent. 8 respondents, however, specified that the tribals are instigated by outsiders to do most of these things.

Yet the general opinion amongst the respondents was that the tribals misuse the concessions given to them.

Abolition of Forest Villages

When asked why were the forest villages converted into revenue village, 9 out of 10 respondents answered that this was done to bring the residents of forest villages in the mainstream of development. Earlier they were not covered under the 'tribal development and welfare' schemes run by the Revenue Department. After being transferred to the revenue department, they are entitled to the benefits of these schemes, one respondent differed from this opinion and said that the decision of this conversion was purely a political decision. Once lands are transferred in the name of the tribals, as is done by transferring village to the revenue department, outsiders can easily buy it or somehow encroach upon it.

Nine respondents were in favour of these conversions and one respondent was against it for the reasons given above.

Next question was asked to know if there were any encroachments at the time of this conversion. Six respondents accepted that there were encroachments done by local people and also by landless outsiders from Madhya Pradesh, Gujarat and Andhra Pradesh. Four respondents disclaimed that any encroachment were made at the time of converting forest villages into revenue villages.

Management of Forests

'Do you think it is possible or desirable to associate tribals with forest management' was the next question.

Out of the 10 respondents, eight answered in affirmative and added that the tribals can do so by helping in guarding the forests, protecting them from fire, stopping illegal feeling, plantation, prevention of deforestation, collection on minor forest produce etc. two respondents felt that tribals have no role to play in the management of forests. But the majority felt that tribals can definitely help in forest management.

When asked as to how the tribals can be given a stake in forest development, respondents suggested schemes like social security plantations, tribal ownership of forests, priority to tribal needs of forest produce and policy of administrative prices. However, two respondents who did not feel that tribals can be associated with the management of forests, did not answer.

Next question referred to ways in which tribals can be a hindrance in forest management. Six respondents felt that tribals were not a hindrance at all. Remaining four, who felt that the tribals were creating problems in forest management, pointed out some ill-practices adopted by the tribals. For example, encroachment on forest land, girdling, grazing in plantations, illicit felling etc. But the majority of the respondents did not consider tribals to be a hindrance in forest management.

Illegal Felling

What is the tribals' share of illegal felling, to the total, was the next question, four respondents said that the share was insignificant. Six respondents felt that the tribals share is as high as 80 to 90 per cent to the total. These respondents related it to the high rate of unemployment in forest areas.

Regarding the statement of the National Commission on Agriculture that, "headloads are converted into cartloads, and cartloads into truckloads, truckloads into wagon loads," four respondents felt that it was an over statement of the facts. Six respondents said that though it was partly true, the tribals are instigated by the urban people. They also added that the truck owners and wagon loaders were mainly local politicians, saw-mill owners and contractors.

The question asking as to who finally benefits from such illegal activities, all the ten respondents accepted that only non-tribals, middlemen and manipulators benefitted though the tribals also had their share in it.

Biased Treatment to Tribals

Undue concessions given to paper mills was the next question presented through quoting an instance where a paper mill in Karnataka was given bamboo at the rate of Re. 1 per tonne. Eight respondents said that it was purely a political decision and an unjust decision. They also added that previously bamboo and other forest produce was available in abundance, and therefore, had no value. Two respondents were not aware of such concessions.

Asked if such bias was shown in their respective circles, four respondents quoted similar examples from their circles. Remaining six respondents were not able to quote examples.

Employment

The question that why do the tribals need to sell head loads illegally and why does the Forest Department not provide alternate employment, seven respondents felt that the main reasons for this was the seasonal nature of employment provided by the Forest Department. Due to this tribals have to get involved in the risky jobs like illegal sale of fuelwood. Remaining three respondents referred mainly to the low standard of living and poverty of the tribals. In general all the respondents agreed that the tribals are forced to indulge in the illegal dealings and do not do it willingly.

All the 10 respondents agreed that the nature of employment created by the Forest Department is seasonal in nature. The department employs labourers for not more than six months in a year.

The question as to what can be done to make forestry employment more regular, three respondents straightway said that nothing can be done about it. Seven respondents felt that it can be improved with proper planning of the forestry work.

The question as to whether there is mechanisation in forestry, all the 10 respondents felt that the mechanisation has been only marginal (nominal). Two respondents felt that this mechanisation would result into decreasing employment opportunities for the tribals, increasing demand for skilled labourers and increasing rate of destruction of forests. Remaining eight respondents felt that no such change will necessarily result in.

When asked if forestry wages were better than the agricultural wages in their respective areas, eight officers said that forestry wages were better. However, two respondents accepted that, in their respective circles, agricultural wages were at par with forestry wages. But in general, majority of the respondents felt that forestry wages were higher than agricultural wages.

Forest Development Corporation

It was mentioned that as a result of the present policy of the government FDCM is felling the natural forests and planting commercial species. The question was regarding the effect of this policy on ecology, industries and the tribal needs.

Six respondents felt that the commercialisation would affect ecological balance adversely. Other four felt that it would not make any difference.

Regarding industrial needs, eight respondents felt that the policy would affect industries favourably, while two felt that it would not do so.

Eight respondents felt that the policy of clearfelling natural forests would not affect the tribals at all because tribal needs can be fulfilled by remaining natural forests. They also substantiated their argument by adding that only a small percentage of total forest area is given away to FDCM for clearfelling. Only two respondents emphatically said that this policy will adversely affect the tribals, because the area earmarked for FDCM is a permanent loss to the tribals.

The question on advantages and disadvantages of the 'food-package' programme undertaken by the FDCM was answered only by four respondents. Others seemed to be unaware of it. All the four respondents said that it was a good incentive for the tribals to come for work, and also a method of providing food at low price. The only disadvantage mentioned was the increasing dependence of the tribals on the 'food-packets'.

Social Forestry

The question was, in how many years will the social forestry be able to take care of the fuel and food needs of the people? It was answered as 'never' by three respondents. They added that it is so because social forestry needs co-operation of the people which is not there. Other seven respondents found it equally difficult to say anything for sure.

When asked as to why do the tribals lack interest in various social forestry projects, 8 respondents felt that it was lethargy of the tribals which kept them away from such projects. Two respondents added that the defects in the organisational set-up, and the unprofitability of the project also are equally important reasons.

Priorities in the National Forest Policy

The respondents were asked to rank the priorities as given in the national forest policy, to ecology, industrial needs, and tribal needs.

Seven out of 10 respondents felt that forest policy has given top priority to the ecological need. Two respondents said that industrial needs are given top priority. Only one respondent felt that tribal needs are given first priority in the forest policy.

Seven respondents gave second rank to the industrial needs while three respondents gave second rank to the ecological needs.

As many as nine respondents accepted that the policy gave minimum importance to the tribals' needs. They also added that it is so because the tribals look after themselves well and easily get whatever is necessary, from the forests. Only respondent gave the third ranking to the industrial needs.

Collection of Fuelwood

It was asked as to who is responsible for the situation where tribal women and children have to walk for miles to collect the fuelwood daily. Four respondents felt that this was an over statement of facts. Even today tribals can get all the necessary fuelwood within their easy reach. Four opined that tribals themselves are responsible for this as they have been over exploiting the forest produce available near their habitats. Only two respondents felt that the policy-formulators and contractors are responsible for this.

Plantation and Felling

Next two questions were regarding the area under tree-cover and rate of clear-felling in the respective circles of the respondents.

Eight out of 10 respondents accepted that the area under tree cover in their circles has been going down. Two respondents, however, did not agree with this.

Seven respondents also accepted that the rate of clearfelling is more than the rate of plantation in their respective circles. Three respondents felt that this is not true.

Alternatives to Timber

All the respondents accepted the need to develop alternatives to timber. They suggested that there should be increasing use of plastic, steel and other metals, fibre-glass, plywood, concrete, sugarcane remains and agricultural residue for paper etc.

General

Regarding tribal movements, especially the Naxalite movement in Chandrapur and Gadchiroli, seven respondents felt that these would bring no good to tribals, and therefore, tribals should not get associated with them. One respondent, however, said that these movements were well organised by educated and dedicated workers, but could not say anything about their effects on tribals. Two respondents did not answer the question.

The question on effects of the displacement due to irrigation projects on tribals was answered by five respondents only. The remaining five, were not aware of the problem probably because no displacement had taken place in their jurisdiction. Four respondents said that this kind of displacement dislocates the tribal economy and culture affecting them adversely. Only one respondent said that it affects the tribals favourably as they get good compensation in the form of alternate land, as well as cash.

Inferences

Following inferences can be drawn from the interviews of the 10 forest officials —

(1) In general, forest department pays inadequate attention to the tribal problems and majority of them consider tribals as forest labourers only.

(2) The forest department considers tribals only in one role, i.e. as

destroyers of forests, unaware of the problems faced by the tribals due to reducing availability of forest produce.

(3) Privileges and concessions given to tribals are considered as favours and also a burden. The department considers itself as a protector, protecting the forests from the misuse of intruders — the tribals.

(4) Some officials working in forests are not aware of the ideological considerations behind certain change such as conversion of forest villages into revenue villages, declaration of Reserved forests, acquisition of private forests etc. For most of them, increased production of forest produce is the only goal of forest policy.

(5) Many of the respondents appeared to be unsympathetic towards the sufferings of the poor tribals, and were as if, biased against them. The problems of encroachments, illicit felling, collection of minor forest produce, grazing of cattle in forest lands were looked at without considering the root cause of the evil. Tribals were rather quickly termed as intruders and enemies of forests.

(6) Two respondents were against associating tribals with forestry development, because in their opinion, tribals were lazy and incompetent. The aspect of traditional know-how of the tribal communities was totally ignored by the respondents.

(7) The tribal participation in organised illegal felling was highlighted, but very few appreciated that the poor are compelled to do so for their livelihood, while the actual profits are pocketed by traders and even local politicians only.

(8) It could be observed that not much thought was being given to increasing the employment opportunities for tribals, and making forestry employment regular and more remunerative.

(9) The failure of social forestry project was also attributed to the lethargy of tribals and only few showed awareness of the inherent drawbacks of the project from the point of view of tribals, namely, its unprofitability etc.

(10) On the question regarding fuelwood crisis, growing industrial pressure, and disturbed ecological balance, it was found that usually the tribals are held responsible for this State. It is forgotten that before the so called 'scientific forest management', these problems had never arisen.

(11) Despite the fact that mainly outsiders are responsible for

encroachments of forest lands, the blame is usually put on the tribals.

(12) In general, it was observed that like the forest policy, the respondents also think very little about the tribals. Their work is related to the forests, but not to the people who live in forests. Tribals are deemed to be good as long as they quietly obey the officials. But their customs, their dependence on forests, their need for forest produce etc. are all treated as liabilities.

(13) Tribals knowledge of forests, trees, herbs, different uses of trees, flowers and fruits etc., are considered as almost unscientific. The trees excepting teak, eucalyptus and the like are considered worthless.

(14) Schemes like 'cheap food packages' under the UN funded World Food Aid programmes are considered as tribal welfare programmes. In such schemes proper thought is not given to the eating habits of the tribals and the consequences of the withdrawal of such an aid, on the tribals.

SECTION III

SAMPLE OF TYPICAL VILLAGES

As mentioned earlier, originally this sample was not a part of the scheme. At the time of the field-survey of the forest villages, it was felt that the 'impact of the forest policy' could be highlighted with the help of some typical villages. Five such villages were selected, namely, an unauthorised settlement, an expropriëtary forest village, a village under the Tribal Development Corporation of Maharashtra, a Kathiawadi settlement, and an irrigation-project settlement created by the Revenue Department. Questions regarding the background of their typical nature, the problems faced by the residents, and the role of government's policy, were put to the residents. The observations made with the help of such informal interviews are compiled for each type of village separately.

An Unauthorised Settlement

Village Kalibel is presently situated in Taloda range of Mewasi Division in Dhule Circle. It is not to be found in revenue records as it is an unauthorised settlement.

The residents of Kalibel migrated from Chulwad, a village in Dhadgaon taluka, some 10 to 15 years ago. The main reason for this migration was that in Dhadgaon, cultivation was done mainly on hill-slopes which became unproductive due to heavy soil erosion as a result of large scale denundation of forests.

These cultivators were settled in Kalibel by some local political leaders after charging Rs. 8 to Rs. 10 thousand each. Those who settled, had to clear-fell 2 to 3 acres of area allotted to them by the leaders. In the process, the political leaders must have made thousands, if not lakhs of rupees. There are 100 such families living presently. It is obvious that they could not have done all this without coming to the notice the 'forest guards' or other employees of the forest department.

Due to the lack of irrigation and other facilities, the productivity of the occupied land is low. In the off-seasons the main source of income for the residents of Kalibel is illicit felling and sale of head-loads in the local markets.

Forest Department tried to evict the residents of this village once in 1982. The villagers fought their case up to the Supreme Court which stayed the eviction in 1983. Since then the villagers are regularly taking crops from the encroached lands.

Maharashtra Times, a Marathi daily, published from Bombay, had (13th March, 1988) published a well studied article on the eviction of unauthorised settlements in Dhule circle. The article has quoted examples where families are living for generations in so called unauthorised villages. The forest officials are reported to have accepted their inability to take any legal action against such settlements because of the 'stay order' issued by the Supreme Court. But when asked as to why no action was taken before the court's stay order, the officials referred to some political interferences. But as the author sees it, "the real reason for inaction is the red tapism and corruption in the forest department".

The forest policy is very strict regarding the use of forest produce by tribals, grazing of tribal cattle, cultivation on forest lands etc. But the example of settlement like 'Kalibel; brings out loopholes in the policy.

An Ex-Proprietary Forest Village

Dab is the name of a village in Dhule Circle, which was under the ownership of Private forest chieftains till 1975. Though the village was handed over to the Revenue Department under the Private Forests Acquisition Act, 1975, and was never a 'forest village', it is very much located in the forests. It also has supplied labour for forestry works since long.

After witnessing large-scale encroachments on the foot-hills of Satpuda in Dhule circle, one climbs the hills in the hope of seeing really thick mountain forest. After to hours of tedious journey on tough hilly roads, which can be covered only by jeep, one is really disappointed to find absolutely denuded forests — if they at all can be called forests. With just one of two trees within the eye-range, it is impossible to believe that this area was covered with thick green forest only few years ago. Chieftains, the owners of private forests underestimated the wealth they owned, and leased out the forests to contractors at very low rates. Though they did not maintain proper supervision on the working of the contractors, Forest Department had employed Range Forest Officers (RFO) to supervise the task. The RFOs were to ensure that only (marked) matured trees were felled. Today's disdainful picture clearly narrated the quality of supervision (amounting to active collaboration with contractors !) done by the concerned forest officials.

Dab village is also typical in another sense. It is scattered in 14 *padas* (clusters) — each pada containing 2 to 3 houses. The villagers own 4-6 hectares of land per family, and yet have to purchase fuelwood, and fodder for their cattle. They are in the forest area only nominally. They can not avail of any concessions and privileges because there is no forest produce available. Encroachments are so common and substantial that it is nearly impossible for the forest employees to differentiate between the owned and the encroached land.

Thus, the objective of the forest policy which emphasised the need of maintaining minimum of 33.33 per cent land under tree cover has patently failed.

Village under the Tribal Development Corporation of Maharashtra

About 10 km from Bhamragrh in the South Chandrapur circle, there is a village named Piarkoti, it was settled in 1964 by the Revenue Department under tribal development scheme and is presently looked after by the Tribal Development Corporation of

Maharashtra. The village has been inhabitated by 'Madias' who were previously scattered over the hills. These tribals were settled on plains to discourage shifting cultivation on hills practised by the villagers, and also to develop them. The population of the village is 150, with 27 families. The Corporation has given 5 acres of land for cultivation to each family..

Even after 23 years of settlement, it was found to be the most backward village as compared to the villages in the sample. The villagers were also given a pair of bullocks each, which was sold off. It was reported that no Tribal Development Officer had visited this village for years. The resident tribals subsisted on rice, bamboo shoots and mahua flowers. The Forest Department neglected the village as it was under the jurisdiction of the Tribal Development Corporation. The village people said that none of them was employed in any forestry work. The sufferers were the poor tribals who did not even know what is being done in the name of their development.

A Kathiawadi Settlement

The lack of co ordination between two governmental departments, working in the same area, was also apparent in a Kathiawadi settlement in Dhule Circle. This settlement is also an example of how one governmental department, in the absence of information, works against the interest of other departments.

On the Bombay-Agra highway, 110 km away from Dhule, there is a village named Panakhed situated in North Dhule division of the Dhule circle. It is a typical village of Kathiawadi settlers. These Kathiawadis have migrated from Gujarat in large numbers mainly because forests in Gujarat are reportedly completely denuded, and grazing is strictly prohibited. These Kathiawadis own a very large number of cattleheads — on an average 40 to 50 bullocks and cows per family. They sell milk for their livelihood and do not own any land for cultivation.

Because of the large size of their cattle herds, they are considered as great hazard or as 'cancer' by the forest officials. Kathiawadis' main problem is fodder for their cattle, and to this end, they fell *Anjan* and other good fodder trees, or let their cattle graze in plantations. This is harmful to new plantations on the one hand and adversely affects the fodder supply to the local forest dewllers on the other.

For all these reasons, the coming of Kathiawadis is disliked and resisted by local villagers as well as forest officials. But the settlement of *Panakhed* villagers was reportedly encouraged by the 'Dairy Development Corporation' of Maharashtra to facilitate the purchase of cheap milk. Nothing more was done for the development of the village in the form of grazing facilities or cheap fodder etc. Presently the villagers supply cheap milk to one governmental agency, i.e. the Dairy Development Corporation, but are a nuisance to the other, i.e. the Forest Department.

An Irrigation Project Settlement created by the Revenue Department

Residents of village Zende Anjan were rehabilitated by the Revenue Department because of an irrigation project. These residents are originally from a village called Dukripada in Sakri taluka of Dhule district, where an irrigation project was undertaken on the river Kann. It is noteworthy that these settlers came from forest area.

The villagers were given alternate land which was less in area, than their own land. The alternate land was also undeveloped. Therefore, the compensatory money given to them was all spent on making the land cultivable. The villagers now face problems which were unknown to them in Dukripada. Here they face scarcity of potable water, fuel and fodder. Because of the highly denuded state of forests surrounding the area, the villagers cannot gather any forest produce — which was adequately available in their earlier village. They have lost contact with their relatives who are left behind, and are now being treated as aliens by nearby villagers. This cultural and emotional break cannot be compensated in any manner. But the most sarcastic fact is that the villagers who were uprooted for construction of an irrigation project are not provided with any irrigation facility in their new settlement. They complained about the lower productivity of the new land due to lack of irrigation facility.

Such drawbacks, loopholes in the government's policy affect the rural poor who do not know how to fight for their rights. What is left for them is only to suffer and blame their own fate. One reaches a painful and inevitable conclusion that the government which claims credit for the development of these poor displaced cultivators is, in reality, responsible for their continued plight.

Inferences

Some broad inferences are drawn here on the basis of the study of five typical villages:

(1) The large scale felling of the forests has resulted into soil-erosion on mountain slopes. This has substantially reduced by productivity of cultivation done on hill slopes. Because of this tribals are forced to migrate to other places, usually plains, for cultivation.

(2) There are loopholes in the present forest policy which is very strict against grazing in forests, collection of forest produce etc. on the one hand, but allows large unauthorised settlements on forest land for political reasons, on the other.

(3) Though it is a policy decision that only old and matured trees should be felled, lack of supervision and other similar reasons have resulted in illegal felling on large scale.

(4) In many places there are privilege holders only for name sake. There is no forest produce available in forests for consumption.

(5) There is a lack of cooperation between two governmental departments working in same area. Therefore, many times, one department works against the interest of the other.

(6) Giving a piece of land and a pair of bullock is considered as tribal development. The follow-up which is necessary is often neglected.

(7) The policy of rehabilitating the displaced is full of drawbacks. The people who are evacuated for an irrigation project are settled without assuring irrigation facilities to them.

(8) The government which claims credit for the development of these poor displaced cultivators is, in reality, responsible for their continued plight.

Summary

In this chapter the impact of the forest policy on tribals was discussed on the basis of field information and observation.

In Section I, the study of demographic and economic aspects of the sample forest villages reveals that tribals usually have large families and are generally illiterate. Dwindling forest cover has severely affected tribal dependence on forests adversely affecting their standard of living. Insufficient employment and low opportunities of self-employment have forced the tribals to indulge in illegal activities like encroaching forest lands, illicit felling, selling forest produce to finance self-consumption etc. At the same time, the revenue-oriented forest policy of the government, by its single-minded pursuit of commercial plantations, has grossly neglected the needs of tribals.

Section II has dealt with the reactions of the employees of the forest department, working in and for forests. It seems that many forest officials are not aware of the tribal problems, and consider the forest dwellers only as labourers for forestry works. The privileges and concessions enjoyed by the tribals are considered as burden on forests, but the undue concessions to forest-industries are ignored as 'political decisions'. Tribals are held responsible for the present state of degraded forests, forgetting that before the 'scientific forestry' was introduced in India by the Britishers, this problem had not arisen.

Section III has considered some typical villages. The study of these villages has revealed that many times there is no coordination between various governmental departments. This results in increasing the sufferings of the tribals living in such villages. Faulty policy, wrong implementation of certain policy decisions, and defective management, have resulted in decreasing forest cover on the one hand, and eroded standards of living of tribals, on the other.

PART III

6

CONTOURS OF LONG-TERM RELATIONSHIP

The forest policy of the Government of India, adopted right from the pre-Independence period, can be stated to be short-sighed. It resulted in denudation of forests, disturbing the ecological balance, causing inadequate supply of raw materials to the forest-based industries, and it also uprooted the culture and the economic means of sustenance of the tribals. This chapter is an attempt to visualise a possibility of developing forests through, and along with the tribal communities. This will ensure required tree cover, and at the same time, not only assure supply of raw materials to the industries, but also make it profitable for the tribals to live in and work for forests.

The Relationship

Tribal life is intimately connected with forests. This relationship is always recognised but not defined and interpreted in terms of clear-cut policies and programmes. As a result, tribal economy and forest economy have tended to drift apart with adverse implications on both. In reality, the loss to the national economy has been colossal as forests have declined tremendously. The weakest group, that of tribals, is also severely affected by the scarcity of forest produce.

One of the important aspects of 'man-forest' system is the type of relationship between the two. 'Relationship', in general, is defined mostly in legal terms. But, the 'tribal-forest' relationship stands apart due to its socio-economic and cultural dimension. The tribal communities have evolved in forests through various stages over the time. There are three distinct concepts of ownership of forests which underline this 'tribal-forest' relationship, according to B.D. Sharma (1978).

(1) *Communal Ownership*

In the early stages of human development, when man was a part of the natural eco-system, the communities owned the land as well as forest resources. "The concept of ownership was vague and its effectiveness depended on the man's natural superiority on other users of forests. With the growing pressure of population, the forest area got demarcated between different communities".[1]

With the emergence of the modern economies, forests became an important source of revenue and wealth to the State. Initially, the administration began to exploit the forest resources for commercial purposes, without challenging the rights of the local communities or asserting State's exclusive rights. So long as there was abundance of the resource, tribal communities also did not find it necessary to resist.

(2) *State Ownership*

The situation changed when the administration began to have an assessment of the forest resources in 'economic' terms in a long-time perspective. The local practices came to be viewed as 'destructive'. The State began to assert its rights while tribals still believed that the forests belonged to them. It is at this stage that the conflict arose, and since then the State has been asserting its rights through legal sanctions.

(3) *Corporate Ownership*

In recent years, a new concept of ownership is evolving in order to attract institutional finance for intensive exploitation of forest resources. The forests are viewed as a resource to be owned by a corporate body, public or private, which is to be governed by the considerations of the profitability of plantation. This attitude is, therefore, unconcerned with the local economy. In such a system, the tribals may be supplied with their requirement for a price, may be even at a concessional price for sometime to start with. But gradually this consideration becomes irrelevant and irritant because the new system recognises the 'tribal-forest' relationship in a limited and formal manner only. In this ownership, commercial use of forests is dominant while the consideration of people in the forests is only incidental.

In this way, with the help of sophisticated tools of legislation (as discussed in Chapter 2), the process of systematic dispossession is being conducted over the past two centuries not only by the alien government, but also by the national government. This change-over from the de facto ownership of the forests by the tribal communities to a stage where forest resources are proposed to be worked by private sector on input-output considerations, is a result of the economic policies of the government, dominated by urban and industrial class. This goes against the principle of equal justice and reason; as Marx observed, ". . . the principle of class interest cannot form the basis of the State, for classes represented private interest and thus the State became 'the instrument of private property' contrary to the principle of reason and justice".[2]

Difference in Perception

The conceptual difference regarding forests and forest-produce arises from the fact that forest-based tribal economy, agricultural economy, and urban economy represent three different stages in human development. Perceptions about the forest resources by the tribal communities in the more backward areas and the policy formulators in the government, are vastly different. The fact that a tribal community in these areas can live with 'little efforts' is highlighted by various government commissions and also by the forest department. But their struggle for existence is underestimated or totally disregarded. A conclusion is drawn that tribals are lazy and lack any wish to work for development.

Since the forest produce is treated as a social property, State stakes its full claim over it, and rights of tribals are brushed aside by declaring that '. . . no section of the community can be allowed to wastefully use national economic resources only because it, by accident, happens to live around forests'.[3] It is because of this attitude that the tribals develop animosity for the government and are forced to agitate for their survival and economic justice.

Development of Forests and the Surrounding People — A Synthesis

Looked from a long-term point of view, it appears that the forests and tribals cannot be developed independently. The nation is already suffering from the consequences of a policy followed with zest by an alien government, in which the tribals needs are treated as secondary. The conversion of the local (natural) forest-mix into a fast-growing timber species result only in increase in the revenue for the government at the cost of the people who live in the forests.

At the same time it would be wrong to propose a mere tribal development with no regard for the ecological balance. The needs of the industrial and urban sector and revenue considerations for the government must also be given due place in the long term balanced forest policy. A synthesis is required for fulfilling all these objectives. But it has to be underlined that any failure to adequately and satisfactorily protect the economic interests of the tribal people within their geographical setting by callous bureaucratic thinking and dubbing them as intruders in forest, would lead to two probable consequences. First, for want of adequate work and satisfactory level of income, the tribals would move away from forests. Since agricultural sector is not that attractive, they would move to the nearest available cities, throng the slums and aggravate the problems of urban poverty (on the line of Dandekar-Rath hypothesis on urban poverty). Second, if majority of the tribals are not able to migrate to cities for various reasons, there will be greater tensions within the tribal-agricultural sector. And always dismissing such government policy-generated tensions as creation of Naxalities (as is being done in the case of Gadchiroli) will not help, in the long run. The growing education, it may incidentally be mentioned, is at the same time educating the people and arousing their consciousness about their fundamental right, i.e. the right to survival with the human dignity equal to non-tribal sections of the society. The forest department bureaucracy may, however not like this change in the attitudes of the tribals and may resent the situation slipping out of their traditional unchallenged authority and control.

It has to be ensured, therefore that the local community gets a say in the utilisation of the forest resources, and its economy also gets strengthened by the inputs of forestry development. The path of development should be gradual, and any external inputs of technology or entrepreneurial skills should be in addition to the full utilisation and development of the skills of the forest dwellers. Such a model of development ought to have only minimum of external inputs. It should also be carefully harmonised with the industrialised urban spheres of the economy, giving the interest of the local community equal consideration.

If tribal development and forestry are accepted as two equal goals, it becomes necessary to reformulate certain important concepts. Some of these are discussed below.

(1) *Minor Forest Produce (MFP)*

The tribal economy has subsisted on MFP since time immemorial. In a subsistence economy, where cash requirements of individuals are little, the significance of forest-produce such as fuelwood, fodder, fruits and roots, is very high. The opening up of forest areas to outsiders indiscriminately, has brought about qualitative change for tribals. It has made MFP an important source of income for the outsiders and the government, and has adversely affected the free domestic use of the MFP by the tribals. This situation needs to be changed and reformulate by taking a policy decision to ensure full return of MFP to tribals.

So far, the tribals have been only passive participants in the 'production' of MFP, appearing on the scene only at the collection point. A positive policy should be evolved for ownership and active participation of tribals in growing the MFP so that they become 'growers' and 'owners' of MFP by gradually taking up even cultivation of MFP like dry fruits, tendu, lac, medicinal herbs etc. This would avoid the destruction of forests due to over-exploitation, over-grazing, inadequate care etc. It would also harmonise the interests of the tribal community with the forest economy.

This reform can change the nature of participation of tribals regarding the development of minor forest produce. MFP can become a source or employment and a permanent source of income for the tribals. This will ensure the continued supply of MFP on the one hand, and will make the tribals economically viable, on the other.

(2) *Ownership and Rights*

Divergent claims about the ownership of forest resources is a crucial and controversial issue. The forest department tends to view the tribals as intruders and encroachers while the tribals see the forest department as exploiters of 'their' forests.

At present, little effort is being made to take the tribal communities into confidence, persuade them to become tenderers of trees and follow scientific exploitation. A situation needs to be created where tribals can accept and create an extensive and better tree-cover in their own interest. They should be given a stake, as an individual or as community, in the tree wealth. They should be associated in the programme of tree-plantation. They should be given tenancy/ ownership rights at least on trees which they help to grow if not on lands so that they can be expected to look forward to the groves as a source of their sustenance.

This will mean a change in the relationship between the State and the tribals. Instead of employing tribal labour for plantation and other forestry works, forest departments can become guardians of trees, and helpers of tribals.

(3) *Fuel and Fodder*

Other important tribal needs of fuelwood and fodder are directly fulfilled by forests. Improved varieties of grass should be introduced to increase its supply. According to a study undertaken by Nimbkar Agriculture Research Institute at Phaltan, wasteland should be used to grow fodder. The study has emphasised that the survival rate of fruit-trees planted in bad quality soil, is nearly zero. It has further shown that on lands which have not more than 30 to 40 c.m. of top-soil layer, the growth of trees is retarded and only one tonne timber is produced per hectare per annum. Instead, if the wastelands are used for planting grass, the loss of rain water decreases from 85-90 per cent to 30 to 40 per cent. Soil erosion decreases from nine tonnes per hectare to two tonnes per hectare. The study has further calculated the increase in the production of fodder which can result in increase in sheep or goat mutton, wool from sheeps etc.[4] Therefore, improved varieties of grass should be introduced to increase the production of fodder and the utility of wasteland.

In the proposed planned utilisation of forests, suitable sections should be identified for the grasses on such a large scale as to become fodder depots. Organised on modern scientific lines, the fodders may be supplied not only to tribals but after getting properly processed and packed, could be supplied to the nearby rural and also the urban sector. This can surely increase employment. This can be undertaken on departmental basis or by the FDCM. The establishment of fuel and fodder depots within easy reach of tribal settlements, ensuring the supply would be of great help to the tribal communities.

Just as in urban development planning 'green belts' are introduced and maintained along with the 'yellow-belts', fodder and fuel-belts can be created around the tribal habitations. New experiments need to be undertaken to produce low-priced, nutritionally rich, and environmentally healthy cattle fodder. Similarly, fuelwood plantations need to be planned on a long-range basis and replenishment of exhausted stocks should be undertaken on a routine basis.

Similar belts can also be prepared for medicinal herbs, new varieties of fruits and other minor forest produce etc. which can give quick, regular and as much income as the timber can give. If the produce is disposed off through the tribal cooperatives, ways can be found out to satisfy the needs of sustenance of the tribal people.

(4) *Employment*

At present, employment in forestry is mostly seasonal and casual in character. The individual tribal is not sure either about the duration of employment or its periodicity. Also, the nature of employment in forestry is such that it coincides with the timing of agricultural operations. Therefore, though the supply of labour in tribal area is high, forest department has to hire outside labour. It is possible to reschedule the forestry works in such way that the local labour is not denied the opportunity of employment in the forestry works.

Very little attention has been paid to the processing side of the forest produce which is sold in raw form. Processing can be a great source of employment for the tribal communities. It involves some training and supply of tools etc. This can be organised on a cooperative basis. The Report of the Scheduled Area and Scheduled Tribes Commission (1961) has rightly commented that, "There is no reason why major forest produce cannot be processed in or near the forest areas, for this would be cheaper and more meaningful. Export of major forest produce in its raw form not only deprives the tribals of wages but also deprives them of the opportunity to develop".[5] Sawing, polishing and furniture making etc. can create many employment opportunities. It has been observed that saw mills owned by the forest department in some circles are kept idle, even while wages are paid to the staff and the sawing work is enabled to go to the private sector saw mills. This points towards the neglect of government, tribal and peoples interest in general and protection of interests of the private sector in particular.

Even the forestry operations, if well planned on long-term basis, can generate considerable employment of a permanent nature. In the forest departments there is a lot of scope for recruiting some of the tribals as forest guards etc., while others could be trained and utilised on a planned basis for the forestry works like plantation, numbering of trees, nursing of tribal medicine etc.

A point is always made by city-born and brought up officials that tribals are lazy and do not want to work. This point can be replied by saying that with a continuous economic development, security and education for a century or so, the tribals would be as honest and industrious as we claim ourselves to be today. Only they have to be given opportunity to develop. Providing regular and gainful employment is one of the important steps to do so.

(5) *Afforestation*

Any attempts towards afforestation without taking into account the needs of local population will not solve the problems of the tribal people. Trees have been planted traditionally for fruits, roots, fodder and fuelwood. In the pre-British times, rich forests were adequate to meet the needs of the local tribals. Only after heavy deforestation did the question of afforestation arise.

Several afforestation projects have been taken up till now, such as social forestry, plantation of private lands, plantations given on lease etc., but most of them have failed. Some of the important reasons for their failure are :

(i) wrong selection of planting sites and species,
(ii) improper seed collection,
(iii) incorrect nursery and plantation techniques,
(iv) improper supervision of the plantation programme, and
(v) widespread grazing by domestic animals (even in plantations fenced with barbed wires).

In addition to these, another important factor is the lack of consideration for those who live around such plantation. L. Fortman in her article on Land Tenure and Tree Tenure has aptly said that, "A project may be destroyed because it has interfered or has itself destroyed people's traditional use-rights. A project may fail because certain groups of people are not allowed to plant trees or because the planters are not guaranteed the use-rights".[6]

A vested interest needs to be created in people regarding afforestation. At present, the industrialists do not have a vested interest in saving forests because afforestation is an expensive proposition and they can manage to get raw materials, at subsidised rates without themselves investing in long-term tree plantation. The tribal people do not feel that they have a vested interest, because the ownership of the forests has been taken away from them.

If the problem of tribals economic development has to be tackled while conserving the forests, it is important to convert afforestation into an income-generating enterprise. By making necessary changes in the ownership and organisational pattern of forest management, tribals can be encouraged to undertake afforestation which would be more profitable than cultivation. Community rights can be restored in some new form for this purpose. In the long term, with training and development, the growers can be made sharers of the profits from the forest produce. The forest department can be made a sole producer and owner of high-class timber, which is not directly needed by the tribals, but they can be co-opted with production of fodder, fuel and other minor forest produce. With definite planning, species to be grown could be selected to ensure ecological balance, rural needs of food, fodder and fuel, and also the profit generating industrial needs.

(6) *Tribal Development Programme*: (Equation of wage-employment to development)

Usually tribal development is equated with employment generation only. For example, the projects taken under the tribal area sub-plan, do not provide anything but employment on wages to the tribals. Tiger projects, National parks, Forest guard posts etc. are undertaken in the name of tribal development which is in no way directly benefitting the tribals. As a scholar of tribal studies, B.D. Sharma puts it, "One of the important reasons for a comprehensive development programme, taking all aspects into consideration, not emerging so far, has been that mere provision of even casual wage-employment in forestry operations has been treated as a measure of tribal development".[7] Employment generation and wages are a necessary but not a sufficient condition in strengthening the tribal economy in the long run. Sometimes this approach is adopted even in plantation programmes taken up as a part of tribal development schemes. The tribals get the wages for the work done while the permanent assets become the property of the State.

The basic objective of a tribal development programme should be to make each family economically viable within a 'clearly defined time-frame'. Therefore, a programme can be deemed to be a tribal development programme only if it has a potential to strengthen the tribal economy on a permanent basis. The financial support of the government to the tribals, in that cases, has to be used for converting their labour into their asset. A plantation scheme with ownership of the tribal in the tree and its usufruct would qualify as a tribal development programme. Whereas a plantation scheme which uses the tribals as wage-labourers, and add to the capital assets of the State, is a mere forestry development programme and not a tribal development programme.

Several programmes can be thought of which would serve both the purposes — economic development of tribals and restoration of forest wealth, simultaneously. Experiments could be undertaken where 5 to 10 acres of land is leased out to tribals for growing tree cover with the help of technical guidance by the forest departments. Many such programmes have already been undertaken which, can however, be made more successful while many other such schemes can still be formulated.

(7) *Participative Forest Management*

In any scheme of working of forests, the local tribal community which provides the labour, should be accepted as partner in the management and sharing of profits. They should not be taken merely as casual wage labourers whose services can be dispensed with at will by the Department. The forestry labour should be brought under special contractual relationship with the forest management which may take into account the seasonality of forestry operations, their complementary place in the tribals' economy and work schedule, and the need for giving a continuing interest to the tribals as individuals or as groups with well defined rights and obligations. The tribals should also be entitled to long-term benefits on the same lines as labourers in other organised sector are. If this concept is accepted, a healthy partnership can be developed in the working and management of forestry resources in the tribal area resulting in the tribal development and the forestry development programmes reinforcing each other.

Participation in management can be brought about by organising this work-force in a different way, e.g. by forming forest labourers' cooperative societies. If the entire work in a region is entrusted to these societies, they can be made to organise their manpower supply on a voluntary basis. The community can learn to adjust its agricultural work-schedule to the forestry work-schedule. The tribal communities even now have not forgotten collective decision-making. In this management, an individual will be a labourer as well as a participant in policy formulation. This will provide a sound base for continuing relationship between the management and the labour. The profit-sharing can also be introduced on certain mutually agreed principles.

If these arguments are basically accepted, then different working arrangements can be evolved. For example, the local community can be visualised as the supplier of raw materials to the industry (by planting and rearing the necessary species by themselves) on the same lines as producer of raw materials in other industries. If the lease for working of forests can be given to the community or co-operatives of labourers, the arrangements of supply on schedule can be arranged. A working plan for extraction, taking into account agricultural operations and requirements of industry, can be formulated.

It is through the active participation of the community in the decision-making process related to the management of forests, that the tribals can be expected to share the common goals set for the development of the region with forestry as an important element.

Basis of New Policy

It can be argued that tribals and forests can be best developed when developed simultaneously. Therefore, it becomes necessary to formulate a new forest policy with this basic conviction. Tribals should not be considered merely as forest-dwellers but as *adivasis* (oldest dwellers), who have their social, cultural and political origin in forests. They should not be developed detached from forests. They should not be treated as people living around forests merely by chance, and the government being a benevolent landlord, giving some concessions etc. It must be remembered that the tribal communities were the owners of forests. It is only now that the urban society is placing unbearable burden on them. So a policy needs to be worked out in which the tribals are not deprived, but are developed and made equal beneficiaries with the urban population. They should be allowed to remain in forests and develop themselves.

A Balanced Policy

What is required at this stage is a new forest policy that attempts a balance between the tribal, environmental and industrial needs of the forest-product. The objectives of the policy should be well defined with clear emphasis. The objectives can be given as —

(a) meeting the basic needs of the tribals such as fruits, roots, fodder and fuel, assured employment and sustained improvement in the standard of living i.e. their economic development,
(b) protection of the environments and ecological balance, and
(c) needs of the industrial raw materials and needs of urban population.

Government's Responsibility

The forest department, as a branch of government, can gradually be made responsible for tribal development along with forest area development. It should not think in terms of an exclusive obligation to trees or vegetation, but should also accept an obligation to utilise the forests as much as possible from the angle of the economic development of the tribals residing in them. This would of course, require a great change in the organisation objectives and attitudes of the Department and the personnel. The extent to which the forests can be protected from the greed of the 'outsiders' will be a challenge to be jointly faced by the forest department and the organised tribal institutions.

An essential feature of this suggested approach is that the forest department should undertake the responsibility of giving adequate employment to tribals living in and around the forest areas. Even the land belonging to tribals can be developed in collaboration with forest department, and Neo-forests could be developed which would be profitable from the tribals' point of view. It has been very pertinently observed that, ". . . for over a century the forest department has been made to be quasi-commercial department, but never a quasi-social or quasi-welfare department".[8] To emphasise that it should now be made so, is the essence of the present work.

Various other government departments working in forest area, e.g. agriculture, tribal welfare etc. should be brought under and co-ordinated by the forest department. Development of tribal housing, language, medicine, culture etc. also should be given due importance.

Rights and Privileges

Large parts of existing forests should be treated as social forests. Rights and privileges over there forest-lands and produce should be revised, listed and made public in local languages. These rights known as 'nistar rights' are even today noted in government records. But they are not known to the potential beneficiaries. The present efforts to curb these rights need to be reversed. Acceptance of the local people's rights on the nearby forest-produce should, in fact, form the core of a people-oriented forest policy.

Forest Development

The planning of forestry development along with tribal-development requires careful analysis of the resource potential of each region as also of the socio-economic situation of the local communities.

Thus, technological forestry research and forestry planning should be the tools to develop forests. Forests should become depots for fodder, medicinal herbs, fruits, fuelwood etc. to meet the basic needs of the tribal community satisfactorily. Only then, forests can also be developed as industrial raw material depots with the help and active participation of the tribal communities. It would not only provide gainful employment for tribals but would also ensure the supply of raw material to the industries.

Sufficient forest area should be treated as ecological preserves. Felling of only old, matured trees, and collection of grass, fallen twigs should be allowed in such natural forests. Once the demand for major and minor forest produce is fulfilled, the remaining forest areas should be treated as reservoirs of natural vegetation.

Hence, a social, equitable, viable, and development-oriented policy is needed which has a partnership approach towards tribals regarding the utilisation and planning of forests.

Summary

Once the interdependence of tribals and forests is accepted as a matter of policy, it should be possible to develop the tribal-economy and forests simultaneously.

The concepts of minor forests produce, ownership rights of the tribals in forests, fuel and fodder, employment in forestry works etc. will have to be redefined keeping the developmental needs of the tribal communities in focus.

The basis of the new forest policy should be tribal development which would automatically ensure required tree-cover, ecology protection, raw material for industries and revenue for the government, in addition to the socio-cultural development and improved economic sustenance to the long deprived people.

NOTES

1. Sharma, B.D., 'Forests, Tribal Economy and Regional Development', Occasional Papers on Tribal Development, Ministry of Home Affairs, Government of India, New Delhi, 1978 p. 1.
2. McLellan, David, *Marx Before Marxism*, The Macmillan Press Ltd., Great Britain, 1980, p. 96.
3. 'National Forest Policy, 1952', Hundred Years of Indian Forestry, Forest Research Institute, Dehra Dun, p. 330.
4. Nimbkar, B.V., 'Desirability of Developing Pastures on Wastelands', *Maharashtra Times* (Dak Edition) 9.3.88, Bombay (In Marathi).
5. Report of the Scheduled Areas and Scheduled Tribes Commission, 1960-61, Government of India, Manger of Publication, New Delhi, p. 136.
6. Fortman, L., 'Land Tenure, Tree Tenure and the design of Agro-Forestry projects,' p. 1 (Private communication)
7. Sharma, B.D., *op. cit.* p. 15.
8. Quoted by Shrikant, M.L. 'Development of Forest Dwellers (Tribals) and Forests', *Vanyajati*, April, 1982, p. 29.

7

SUMMARY OF CONCLUSIONS AND SUGGESTIONS

This chapter summarises the conclusions drawn from the study of the pre-British, British and post-British forest policies and their economic impact on the forest dwelling tribals in Maharashtra. The conclusions arrived at from the field survey of the sample villages in Maharashtra, too, have been incorporated here. Finally, the suggestions have been made for making the present forest policy tribal oriented, and its implementation better.

CONCLUSIONS

Chapter 1

(1) To a vast number of tribal people, the forest is a home, a livelihood; the very existence.

(2) From times immemorial the tribals have enjoyed freedom of using the forests.

(3) The situation changed when the Britishers started entering the forests in middle of the 19th century.

(4) Being aware of the rapid denudation of forests, the Government of India has substantially stepped up the spending on forest development.

(5) The blame for the existing grim situation of the forests is squarely put on the forest dwelling communities without examining the nature of the use of forest produce by the tribals.

(6) The present study endeavours to examine the on going situation of economic relationship between the tribals, forest resources and the forest policy of the government.

Chapter 2

(7) Man appeared in the world as a part of the 'natural system' depending entirely on forests, and developed into a hunter, food gatherer and a cultivator but still remaining an integral part of the ecological system.

(8) In the pre-British period, forest resources were controlled and fairly managed by the local tribal communities, with prudent utilisation of forest produce.

(9) Community ownership of forests was prevalent in India. It did help the superb preservation of the natural resources, i.e. forests.

(10) The British government introduced the so-called 'scientific management' of forests, which had mainly the commercial purposes in mind.

(11) Conservation was neglected between 17th and 18th centuries. The degradation of forests for the purposes of railway expansion etc. finally awoke the British authorities to the fact that India's forests were not inexhaustible and were getting exhausted.

(12) The reservation of forests was done for the first time in 1806. The first memorandum providing guidelines, restricting the rights of the forest-dwellers was issued in 1855, and the Forest Department was established in 1862.

(13) Forests were considered as suppliers of 'raw material for industries' which brought revenue to the government.

(14) The policy of the government enacted through various legislations restricted the rights of forest-dwellers and converted them into mere privilege holders restricting their free access to forests.

(15) The forest dwellers were termed as intruders in forests and were degraded into labourers.

(16) The forest policy in Independent India has also continued to develop revenue — oriented commercial forestry.

(17) Ecology and industrial needs have been given priority in the policy. The tribals, instead of getting back their original ownership of forests, have remained the privilege holders, a burden on the forests.

(18) Various policy measures, e.g. establishment of Forest Development Corporation etc. substantiate the inference that

commercial interests continue to bc dominated at the cost of economic development of tribals, through vicious planning.

(19) The revised forest policy drafting of 1984, 1986 and 1987 of the government of India are not much different in their outlook towards tribals and their rights on the forest produce.

Chapter 3

(20) The Indian Government have always looked upon the forests mainly as a source of revenue and not as habitat of the tribals.

(21) The economically powerful commercial class has succeeded in pressurising the government to give highest priority to the industrial needs on the forest produce.

(22) The ecologists and social workers have conveyed their concern over the degradation of the forests and the forest policy. They have pointed out the failure of the government to achieve its own objective incorporated in the Forest Policy of 1952 to bring 33.33 per cent of the country's total area under tree-cover.

(23) The real sufferers of the depleting forests are tribals who live in and around forests and depend on forests for their wherewithal.

(24) Various functions of the Forest Department denote the anti-tribal bias of the government. The examples are : the commercial forestry through Forest Development Corporations and plantation of exotic species like eucalyptus under Social Forestry to meet the raw material needs of the industries, the present policy of rehabilitation of the displaced persons and the stringent measures on the tribals concessions on forest produce.

Chapter 4

(25) The forest area in Maharashtra is concentrated mainly in the Western, Eastern and Gondvan belts and comprise about 20.00 per cent of the total geographical area of the State, which has not changed much since 1961 i.e. after the formation of Maharashtra.

(26) The tribal population is 9.19 per cent of total population and is concentrated in the Sahyadri belt, Satpuda belt, and the

Gondvana belt.

(27) The Government of Maharashtra have maintained the erstwhile stringent measures regarding the privileges available to the tribals in the State.

(28) The concessions and privileges on the forest produce in Maharashtra are specified circle-wise and to some extent they differ from circle to circle.

(29) As per the government reports, the utilisation of the forest produce by tribals through their privilege, has never exceeded 2.00 per cent of the total extraction of forest produce by all agencies, as against 50 to 60 per cent by the Forest Department and 40 to 50 per cent by the commercial and industrial enterprises.

Chapter 5

The field visit to four clusters and 23 sample villages with 115 respondents revealed the following facts and facets of the economic life of the tribals:

(30) The basic civic amenities e.g. tar road, post-office, electricity etc. are not available in most of the forest villages.

(31) Tribals usually have large families with more than 6 family members and majority (84%) of them are illiterate.

(32) Dwindling forest cover has severely affected tribals' dependence on forests for food, fuel and fodder, adversely affecting their standard of living.

(33) Insufficient employment and opportunities of self-employment, resulting in lack of sustenance, have forced the tribals to indulge in illegal activities like encroachment on forest lands, illicit felling, sale of forest produce etc.

(34) The government has completely overlooked the fuel, fodder and other needs of tribals by its single minded pursuit of commercial forestry.

In the second sample observations were made on the basis of the interviews of 10 forest officials.

(35) Out of the 10 forest employees, most of them were found to be unaware of the tribal problems and considered them mainly as labourers for forestry works.

(36) Two forest employees held tribal responsible for the present state of degraded forests.

(37) Seven respondents indirectly accepted that the privileges and concessions enjoyed by the tribals are burden on forests.

(38) There appears to be severe lack of coordination between the departments working in forest areas for the development of forests and tribals.

(39) Giving a piece of land and a pair of bullocks is considered as tribal development in itself and the follow-up is not done.

(40) The government which claims credit for the development of these poor displaced cultivators is, in reality, responsible for their continued plight.

Chapter 6

The depleting forest cover has resulted into an exodus of the tribals towards cities. In order to avoid this a long — term relationship between the forests and the tribals has to be established on the following lines:

(41) Forests and tribals are very closely related and one cannot be developed without the other.

(42) Tribals had community ownership of forests for generation and it is recorded that they looked after this resource fairly well.

(43) The new concept of corporate ownership which is emerging presently, considers the forests as a source of income only and neglects the role of forests in the tribal culture and economy.

(44) The age-old knowledge of tribals regarding forestry can be used in the development of forests.

(45) Tribals need to be organised and closely associated with the management of forests which will reduce their illegal activities.

(46) Good tree-cover should be developed so as to benefit the tribals so that they look after the resource with care.

(47) Rather than treating the tribals as minor forest produce collectors, they should be made growers of minor forest produce.

(48) All the basic requirements of fuel and fodder should be first assured to the tribals.

(49) Supply of raw material for forest based industries should be made progressively through the tribals communities residing in the area.

(50) Processing of forest produce and even manufacturing of final products can be taken up closer to the forest area resulting in better employment opportunities for the tribals.

(51) Hence, forests and tribals can be developed only simultaneously.

SUGGESTIONS

Introduction

It can be safely inferred from the conclusions enumerated earlier that a new forest policy should be envisioned which will interweave the interests of the tribals along with the maintenance of ecological balance, in addition to supplying forest based raw materials for urban needs. The policy should ensure all these goals along with the achievement of having 33.33 per cent of the geographical area under tree cover.

The total geographical area of Maharashtra is 3,07,690 sq.km About 69,955 sq.km of this area is classified as forests according to 1988-89 figures. This comprises approximately 19.80 per cent of the total geographical area of the State. According to NRSA figures, however, the actual tree-cover in Maharashtra comes to about 30,400 sq.km i.e. 9.88 per cent of the total geographical area. In order to achieve the goal of having 33.33 per cent of the area under tree cover, the state of Maharashtra should have an actual tree-cover of 1,02,563 sq.km. This suggests that for achieving this, an additional area of 72,163 sq.km should be brought under actual tree-cover.

Some suggestions which, when incorporated in the formulation of forest policy, will help to achieve the above-mentioned objectives are listed below:

A. Tribal Orientation

Local People's Rights in Forests

(1) Accepting of local, people's interests and rights on the nearby forest and forest-produce should form the core of a people oriented forest policy.

Uptil now, the forest policy has neglected the traditional rights of the local people, and this has resulted into indifference and animosity of the local people towards forest staff in particular and the government in general. This has adversely affected the forest people's economic condition. If the forest policy gives legal sanction to tribals' rights over forest produce, it will help towards the improvement of the economic lot of the tribals/forest-dwellers.

Such a provision in the forest policy may be objected to by saying that such rights may result into careless over exploitation of the forest produce. However, this suggestion should be viewed in the totality of other suggestions being proposed here. It should also be remembered that in the pre-British period the forests were managed with atleast average efficiency by the local communities. Once the interests and rights of the tribals are created in the nearby forests, they will take care of the resource in their own interest, because the forest will then become a source of additional income and employment for them.

Due Importance to the Needs of Tribals

(2) The objectives of the forest policy should be to give due importance to the basic needs of the forest-dwelling tribals, such as fruits, roots, flowers, fodder, fuel, fertilisers, medicine, minor forest produce etc. Protection of the environment and ecological balance should also be taken care of. Needs of the industrial raw material and the needs of the urban people should be paid due attention.

Neglecting any of these objectives will result into lopsided development. This point is self-explanatory in the light of past-experience where industrial raw-material was favoured, which resulted into heavy depletion of forests, short supply of MFP to forest-dwellers etc. In brief, one can surmise that, the principle of forestry should be to synthesise revenue earning activities with social responsibilities.

B. Forestry Planning

As stipulated in the Indian Forest Policy, it is imperative that 33.33 per cent of the geographical area is to be brought under tree-cover. This would necessitate a phased programme to be undertaken in

Maharashtra to afforest additional 72,163 sq.km of area. In order to achieve this target in, say, next 15 years, every year about 4811 sq.km of area needs to be afforested. The state has 1,44,010 sq.km of wasteland (i.e. much more than the needed land) which can be used for afforestation.

Purpose-wise Afforestation

(3) The area which is to be forested should be earmarked for three purposes — approximately 1/3 of the area to meet the tribal needs (fruits, fodder fuel, timber etc.) to be known as 'community forests', 1/3 to meet the raw material needs, to be known as 'commercial forests' and 1/3 for maintaining the ecological balance, to be known as 'national forests'. It should be noted that the commercial and community forests will also serve ecological purpose to some extent.

It is possible to raise two points against this suggestion : (a) What is the necessity of reserving 1/3 of the forest area as community forests ? (b) Will the 1/3 area reserved as commercial forests be able to cater to the needs of forest-based industries?

The extent of shortage of fuel and fodder in the State clearly highlights the necessity of reserving 1/3 of forest area for community forests. The community forests will not only provide fuel, fodder and MFP to the local communities only, but also can fulfil the demand of the rural and urban fuel, fodder. In answer to the second objection, it should be noted that presently, only 9.88 per cent (i.e. 30,400 sq.km) of the area of the state is under actual tree-cover and is being exploited by tribals as well as industries. In the new scheme of the things, more forest area gets assigned for industrial needs only (i.e approximately 34,188 sq.km). In addition to this, community forests will also provide industrial raw materials to some extent. The suggestion made above can, therefore be treated as realistic.

Site for Community Forests

(4) The plantation to cater to tribal needs should be taken up around the tribal habitats. Fuel belts, fodder belts, belts of

fruits trees, medicinal herbs etc. should comprise such a community forestry.

This will ensure the supply of forest products needed by the tribals within their easy reach and will improve their material well-being to a considerable extent. The urban society will also benefit from increased supply of fodder, fuel, herbal medicines etc.

Site for Commercial Forests

(5) Surrounding the community forests, there should be belts of commercial forests i.e. forests supplying the industrial needs. This will generate employment opportunities to the local tribals.

Financial Involvement of Industries in Community Forests

(6) Vested interest needs to be created among the industrialists to invest in the growth of commercial forests supplying raw materials to them.

Till now the forest-based industries were made available the raw materials at dirt-cheap rates due to which they over-exploited the resource. This new type of involvement (financial) would ensure optimum utilisation of the raw material.

Selective Felling in National Forests

(7) All the virgin forests of today should be declared/classed as National Forests. These natural forests should be totally protected and clear-felling in these forests needs to be totally banned. Only selective felling should be permitted for essential purposes.

This is necessary because a stage has come where any further degradation of forests would be disastrous to the entire society.

Management of National Forests

(8) Forest Department should be given the responsibility of

plantations for environmental needs in general and for plantations by road-side, canal-banks, river-beds, mountainous regions etc. in particular.

Private organisations and social organisations, especially in urban areas, may also be encouraged to come forward for such plantations. However, keeping in mind the importance, the final responsibility of such plantations should rest solely on the government.

Initiative to be taken by Forest Department

(9) The forest department should take initiative in all the three types of plantation schemes. The department should provide the necessary know-how for cultivation and management of the community and commercial forests to the concerned people.

Processing and Marketing of Forest Produce

(10) A comprehensive programme of identifying various forest products, asserting their potential, and also the feasibility of marketing and technological processing of such products should be formulated.

It should be made a law that atleast the preliminary feasible processing of MFP is done in the vicinity of forests. This will help in retaining economic benefits for the local community. The necessary facilities and training in this regard should be provided by the forest and other departments.

Labour Crops

(11) For the purpose of obtaining labour of the right type for various forestry operations, the tribals residing in an around forests should be trained and organised into permanent labour corps.

This will ensure employment to the tribals on the one hand, and timely and proper supply of labour to the forest department on the other. It will create a situation of genuine interest of tribals in not migrating from forest areas.

Self-Employment Opportunities

(12) While developing community forests it should be seen that self-employment opportunities are also generated, and made available through proper organisation. In this regard the traditional handicrafts of the tribals such as wood-carving, basket-weaving, rope making etc. may be encouraged and supported by remunerative prices.

Synthetic Alternative to Timber

(13) The need of the hour is that forest based industries give up all non-essential use of forest produce. The government should encourage more efficient use and reuse of wood products and provide incentives to develop synthetic alternatives to timber. This will help reduce the pressure on the remaining natural forests.

C. Administration and Management

Ownership of Community Forests

(14) The ownership of community forests should be vested in small, manageable and homogeneous groups in the local community itself.

This will help the local population to take care of the local needs. They should be encouraged to take decisions regarding the selection of species, plantation, distribution of produce according to individual needs, and also of the sale of excess forest produce. For such groups, financial and other incentives can be offered for number of trees made adult, for having fully satisfied fodder, fuel, and timber needs of the group etc.

Advantages of community forests should be vested in village panchayats and all revenue from these forests should go to them. Community forests can also satisfy the fuel, fodder and MFP demands of the rural and urban areas and earn revenue. This revenue may be utilised for the maintenance and protection of the community forests as well as to undertake social welfare activities.

Supply of Forest Produce

(15) The required amount of forest producc such as timber for construction of huts, fodder for cattle, fuelwood for cooking etc. should be supplied through the forest depots to be set-up in the nearby areas. These can be arranged by the village panchayats, the produce may be given at concessional rates fixed by the panchayats themselves.

Role of Forest Department in Community Forests

(16) The role of forest department in relation to community forests should be to help and guide the village panchayats for introducing scientific forestry such as — better techniques of plantation, improved variety of grasses and fuel-species, regeneration, exploitation, processing and marketing.

Listing of the Rights of Forest-dwellers

(17) Rights of the forest-dwellers over the forest land will have to be listed and made public in the languages of the tribals. These rights known as nister rights are even today noted but not known to the people for whom they are meant.

A periodical giving details regarding any additions or deletions of rights and privileges, area and species to be reserved, various tribal-benefit schemes, prescribed punishments for offences and changes thereof etc. should be published in their local languages and circulated in the concerned areas regularly.

Role of Forest Department in Commercial Forests

(18) The forest department should be made responsible for management and administration of these forests, which will mainly supply the raw materials to forest-based industries. The department will have to co-ordinate investments made by various industries, and supply of raw material to them. The department can charge in return of the services rendered.

Selection of Species

(19) The department will be responsible for providing the technical know-how of selection of species, plantation patterns, silvicultural methods, exploitation etc.

Ecology and National Forests

(20) National forests should be consciously developed and maintained to cater to the needs of ecology, to avoid soil erosion on river-beds and hill-slopes etc. All the national forests should be mixed forests with all kinds of species chosen according to the climatic conditions.

Tree-Belts in Urban Areas

(21) The department should develop tree-belts in industrial cities and other urban areas to lessen the extent of pollution.

National Forests — Not a Source of Revenue

(22) These forests should not be considered as a source of revenue. However, the sale proceeds of fallen leaves, fuel, fodder, timber of matured — fallen trees, and fruits etc. should go to the state exchequer.

D. Co-ordination

Several government agencies working in forests and tribal areas should be co-ordinated towards achievement of the common goal. The agencies working presently lack co-ordination and at times work against the interest of each other.

Basic Civic Amenities

(23) The policy of the government should be such that it ensures the provision of the basic civic amenities like roads, primary health centres, post offices, schools, water and electricity supply etc. to the tribal villages through the concerned departments. To facilitate the provision of these amenities

clustering of villages can also be suggested. Usually tribal habitats are spread over a large area and therefore it becomes very expensive and very difficult to develop infrastructure on such a large scale. Tribal communities also settle in small and far-flung groups. Providing facilities such as schools, post-office, primary health centre etc. to each such settlement is also virtually impossible in the present set-up. For this reason, a number of such small habitations can be brought together or rehabilitated and then all the above-mentioned facilities can be provided to them. Presently rehabilitations are done in case of project-affected persons. But this kind of clustering of villages or rehabilitations will be a positive step towards tribal welfare.

Research in Tribal Know-how

(24) Efforts should be made to encourage research in the tribal knowledge of trees and medicinal herbs, their language, culture etc. The forestry programmes should be so organised as to take full advantage of the traditional knowledge and skills of the tribals.

This single step may revive and strengthen the traditional know-how, and benefit the whole society.

Time-schedule for Work

(25) A time schedule of forestry work should be prepared in consultation with the department of agriculture, industries etc. in each region with the intention of providing work to tribals throughout the year.

Alternatives to Fuelwood

(26) There is a tremendous scope for reducing fuelwood requirements by propagation of more efficient *chullas* (stoves), solar cooker and alternative fuel such as biogas. This can be encouraged with the help of various government agencies working in the tribal areas.

Stall Feeding

(27) Similarly, there should be a major attempt to change the animal feeding practices to stall feeding of a smaller number of livestocks with intensive programmes of fodder development. The fodder development programmes should be backed up by genetic improvement of livestock.

Educating Forest Dwellers

(28) The tribal communities should be educated to know better methods of cultivation, plantation and nurturing of fruit trees etc. This will improve per hectare yield, leaving more forest produce for sale.

Finance

(29) The State Government should organise co-operatives, and development corporations with a view to consolidate and develop the economy of the tribals and to discharge prior debts, for purchase of MFP, etc.

Training of Forest Department Personnel

(30) For effective functioning of the forest department the training of the personnel of the department should include courses on tribal dialects, culture, traditions, rural development, rural management and community participation etc.

Presently the forest department personnel consider themselves as guardians of forest, policing or protecting forests from the local community. This attitude needs to be changed. As the forests need to be developed now, with the help of the local people, it is necessary for the department to establish rapport with the local people.

APPENDICES

APPENDIX I

QUESTION SCHEDULE FOR TRIBAL HOUSEHOLDS

(A) Identification

1. Name:
2. Father's/Husband's Name:
3. Age:
4. Tribe:
5. Sub-Tribe:
6. Native Place:
7. Village:
8. Taluka:
9. Composition of Family:

Sl. No.	Relation with respondent	Age	Sex	Education	Present Occupation

Occupation Code: (1) Cultivator, (2) Labour in general, (3) Agricultural Labour, (4) Labour in Forest Department, (5) Business, (6) Unemployed (7) Underage/overage, otherwise unemployable.

Educational Code: (1) Illiterate, (2) Primary, (3) Middle school, (4) High School, (5) Graduate

10. **Structure of House:**

    (a) Area (sq.ft.)————

    (b) Roof : local tiles/Mangalore tiles/bamboo/grass and reed/other.

11. **Present Occupation:**

    (a) Cultivator/Labour/Service/Business/any other.

    (b) Cultivated Land/owned/let-out/fallow/let-in.
    Code for Labour (1) Agricultural Labour, (2) Forest Labour and (3) General Labour.

    (c) Physical Location of Land ————

    (d) Quality of Land
    (L.R. per hectare)

12. Were you ever an owner of land? Yes/No
    If yes: Year ——— Area ———
13. How land was lost? — bondedness/non-repayment of loans/poverty/family dispute/acquired for forest/projects/any other.
14. To whom was it lost? Money-lender/master/relatives/government.
15. Today, how far do you depend on forests for fruits, roots and animals?
16. Can you read or do you know how many rights and concessions are available in your circle?

(B) **Concessions From Forest Department**

1. **Food**

Items	Annual Quantity	Value (Rs.)	Distance from residence

2. Medicine

Items	Annual Quantity	Value (Rs.)	Distance

3. Fuelwood

Items	Annual Quantity	Value (Rs.)	Distance from residence

4. Grazing

Items	Annual Quantity	Value (Rs.)	Distance from residence

5. Any other MFP

Items	Annual Quantity	Value (Rs.)	Distance from residence

17. Has the availability of fruits, leaves, fuelwood etc. declined recently due to commercialisation or forest regulations?
18. Problems in availing concessions : Concessions withdrawn/ information not available/many formalities/difficult formali-

ties/red tapism/corruption/language/Distance/non-availability of forest produce.

19. After Independence how tribal life has changed (Give Ranks)

Good	Bad
Improved wages	Movements restricted in forests.
Medical facility	Lost freedom of cultivation.
Regular employment	Restrictions on grazing.
Improved standard of living	Limited collection of MFP
More concessions	Non-availability of fuelwood
Recognition of tribals place in forests	Forest official endowed with more powers e.g. confiscation of tools and property etc.
Housing facility	Compulsion on working in forests if staying in forest villages.
Any other	Any other

20. Your relations with Foresters and Guards: Friendly/Master-Servants/Enemy.
21. Has there been any encroachment on forest land in your village? Yes/No No. of encroachments
22. What action was taken by Government? Immediate removal/refusal to regularise but no immediate action/regularised/staff had to be bribed/any other.
23. Have you ever bribed any Forest Department staff? Yes/No By how much————
24. Has your cultivated land been declared as forest in the past? Yes/No Years————
25. Have any land in your village, been declared forest? Yes/No Year————
26. When your village was a forest village —
    (a) How many members from your family worked for Forest Department ————
    (b) Duration ————
    (c) Wage Rate ————

27. After it was declared a Revenue Village —
    (a) How many worked ——————
    (b) Duration——————
    (c) Wage Rate ——————
28. Are your wages at par with the agricultural wages in surrounding area?
    Yes/No/Does not know
29. Did you prefer having your own land and live in a Revenue village than work with the Forest Department and live in a Forest Village?

(C) **Felling, Illegal Sale**

30. How has illegal felling affected you in general : lack of fuelwood/diminishing grazing lands/scarce supply of fruits and roots/reduced supply of timber, bamboo etc. for housing/low employment opportunities/any other.
31. Who do you think is primarily responsible for illegal felling? government policies/timber traders/contractors/politicians/tribals/any other.
32. How is the fuelwood or other MFP collected illegally? sneaking in at nights/avoiding the forest guards during day time/bribing forest staff/any other.
33. Whom do you sell such illegally collected material? to contractors/to middlemen/to truck drivers/in village markets/any other.
34. How many headloads can be sneaked out at a time? Number ——— Worth Rs. ———
35. How much do you earn per month by participating in illegal felling?

(D) **Projects**

36. There are several social forestry projects. Do you make use of these? Yes/No
    If yes, which projects :——————
37. What are their advantages and disadvantages?

    Advantages : ——————
    Disadvantages : ——————

38. What is your experience with any Social Forestry Project? Useful for employment/can give higher income over a long period/Not much useful for the tribals? / No experience.
39. Have you been benefited by plantation of fast growing species like subabhul for fuelwood? Yes/No
    If yes, the type of benefit : (1) Adequate fuel for family. (2) Adequate fodder for animals (3) Additional income by sale (4) Any other.
40. Do you benefit from mono-culture being emphasised recently?
41. When big Irrigation Projects are undertaken how do they affect your life :
    Give employment/displaces you/affects adversely because of felling at large scale/No experience/ any other.
42. Is the alternate land of equally good quality given? Yes/No
43. At what rate compensation money was given to you?
44. How do you use that money? Creation of assets, repayment of outstanding loans, saving, family expenses
45. What are other effects of displacement on you?
    Disturbs life till all formalities are completed/life-source changes/loss employment/cultivated fields are lost/any other.
46. What are your expectations from Government? Give better employment opportunities/respect their claims on forestry to improve their standard of living/associate them in forest preservation management and planning/any other.
47. Are you a member of any tribal organisation? With which institution is it affiliated?
48. Have you ever reacted against the Government through applications or morchas, representations, court cases etc.?
49. Are your aware of the proposed bill of 1986, its priorities and economic effects on the lives or the tribals?

APPENDIX II

QUESTION SCHEDULE FOR FOREST OFFICIALS

Tribal Needs

1. Has the forest policy of government given necessary consideration to the needs of Tribals?
2. Were the rights of tribals curtailed by British Government to meet their own demands?
3. Has the Government of free India increased its control over forests?
4. Did the British settlement officers do justice to tribals claims?

Community Ownership

5. Was the communal ownership of tribals recognised by the British officers?
6. Whether the tribals were settled in communities or individually?

Restrictions in Reserved Forests

7. In olden days tribals had free access in forests. How, do you feel, the restrictions in reserved and protected forests affected the tribals?
8. After a particular forest area is declared as reserved, what do the tribals do? Leave the place/continue to live and cultivate/any other.

Adverse Effects of Concessions

9. What adverse effects do you observe due to the concessions given to tribals?

Abolition of Forest Villages

10. Why were the forest villages converted into revenue villages?

11. Are you in favour of this conversion?
12. Would you say that there were encroachments at the time of this conversion?
13. To what extent? Who were the encroachers?

Management of Forests

14. Do you think it possible and desirable to associate tribals in forest management?
15. How? By guarding forests/protecting forests from fire/stopping illegal felling/plantation/preventing deforestation/collection of minor forest produce/any other.
16. In your opinion what modifications should be made to give the local communities a stake in the forest development?
17. In what respects do you feel that tribals are hindrance in forest management? By taking more than sanctioned/by trading what is meant for consumption/by false claims/any other.

Illegal Felling

18. Tribals are usually accused of illegal felling, in your opinion what is their share to the total?
19. National Commission on Agriculture says that tribals are the main destroyers of forests saying, "headloads are converted into cartloads, cartloads into truckloads and truckloads into wagon loads". What is your opinion?
20. Who, do you feel, finally benefits?

Biased Treatment to Tribals

21. In contrast to the tribals' pathetic condition, timber traders and paper industries seem to get lot of concessions. How could a paper mill in Karnataka get bamboo at the rate of Re. 1 per tonne while tribals had to pay much more than that?
22. Do you think that such bais was shown in your circle?

Employment

23. Why do the tribals need to sell headloads etc.? Why alternate employment cannot be provided to them by the Forest Department?

24. What is the nature of employment created by the Forest Department?
25. What should be done to make forestry employment more regular?
26. Has there been any mechanisation in forestry works?
27. Will the mechanisation result in decrease in employment/ decrease in wages of unskilled workers/increase in rate of destruction/any other?
28. Are the Forest Department wages at par with agricultural wages in surrounding areas?
29. Present policy of FDCM is clearly giving more importance to commercial forestry. It is clear-felling the natural forests and planting commercially valuable species. How will it affect ecological balance, industrial needs and tribal needs?
30. What are the advantages and disadvantages of the supply of food-packets by FDCM under its World Food Programme?

Social Forestry

31. In how many years will the Social Forestry be able to take care of the fuel and fodder needs of the people?
32. Why do tribals generally lack interest in various social forestry projects?

Priorities in the National Forest Policy

33. Which factors are given priority in our national policy? (a) ecology, (b) industrial wood, (c) fuelwood, (d) tribals needs.

Collection of Fuelwood

34. There have been studies and surveys in which they say that women have to walk miles to collect fuelwood. Who do you think is responsible for this — government policy, tribals, contractors any other?

Plantation and Felling

35. In your circle, can you say that land under tree cover is going down? Yes/No

36. In your circle, is area under felling more than plantation?

Alternatives of Timber

37. In your opinion are there any alternatives to timber?

General

38. What is your opinion about the Naxalite movement in Gadchiroli and other organisations working in your circle?
39. How do irrigation projects affects tribals?

Gupta, R., Banerji, P. and Guleria, A., 'Tribal Unrest and Forestry Management in Bihar', Centre for Management in Agriculture, Indian Institute of Management, Ahmedabad (1981).

Hundred Years of Indian Forestry, Forest Research Institute and College, Dehra Dun (1961).

Huria, V.K. and Acharya, K.T. 'Meeting Basic Needs through Micro-Planning — Central Role of Essential Forestry', *Economic and Political Weekly*, Vol. No. 34, August 20, 1983.

Jha, Prem Shankar, 'Forests and Development : Restoring Primacy to Farming', *Times of India*, April 18, 1983, p.8-35.

——, 'Forest of Interests', *India Today*, November 15, 1986.

Joshi, Gopa, 'Forest Policy and Tribal Development : Problem of Implementation, Ecology and Exploitation', *Social Action*, 31(4), October-December, 1981, p. 446-68.

Kulkarni, S.D., 'Over a Century of Tyranny', *Economic and Political Weekly*, Vol. 9, No. 10, March 9, 1974.

——, 'Law and the Adivasis — Story of the Peasants of Akkalkuwa', *Economic and Political Weekly*, June 28, 1975.

——, 'Alienation of Adivasis Land', *Economic and Political Weekly*, August 31, 1975.

——, 'Encroachments on Forest Lands — The Experience in Maharashtra', *Economic and Political Weekly*, Vol. XIX, No. 45, November 19, 1979.

——, 'Encroachment on Forests — Government versus People', *Economic and Political Weekly*, Vol. 17, No. 3, January 16, 1982.

——, 'Towards a Social Forest Policy', *Economic and Political Weekly*, Vol. XVIII No. 6, February 5, 1983.

——, 'Social Forestry : Need for a Proper Perspective and Policy', *Arth Vijnana*, September 1983, V 25 N3, 289-97.

——, *Conflict and Development, in Population, Poverty and Hope*, Centre for Policy Research, New Delhi, Uppal Publishing House, 1983.

——, 'Adiwasi — Law and Justice', *Economic and Political Weekly*, Vol. XX, No. 28, July 13, 1985, pp. 1171-74.

'Lokrajya', Adiwasi Development Number, December 16, 1983, Government of Maharashtra.

'Maharashtra State Forest Centenary Souvenir', March, 1962, Maharashtra Forest Department, Government of Maharashtra.

'Maharashtra Not Through Forest Lands', *Economic and Political Weekly*, March 21, 1979.

'Maharashtra — Organising the Adivasis', *Economic and Political Weekly*, (Annual Number) February, 1974.

Mahasveta Devi, 'Eucalyptus : Why?', *Economic and Political Weekly*, August 6, 1983.

Meghe, D.R., 'A Deal for Tribes Through Law (Legal Protection to Tribals and Tribal Development)', *Journals of the University of Law*, Nagpur University, Nagpur, 1978-80.

Menon Ramesh, 'Forest Land Fidele', in *India Today*, Vol. XII, No. 13, July, 1-15, 1987.

Moss, R.P. and Morgan, W.B., 'Fuelwood and Rural Energy, Production and Supply in the Humid Tropics, National Resources and the Environment', Series Vol. 4, United Nations University.

Naidu, N.V., 'Scheduled Tribes Exploited by Cooperatives', *Economic and Political Weekly*, May 3, 1975.

National Commission on Agriculture, Interim Report on Production Forestry — Man-Made Forests, August 1972, Government of India, New Delhi.

National Seminar on Social Forestry Through Cooperatives, Department of Rural Development, Vaikuntha Mehtra National Institute of Cooperative Management, Pune, 1984.

'Our Common Future', The World Commission on Environment and Development, Oxford University Press, New York, 1987.

Palit, Sharad, 'Govt.'s War on Adiwasis', *Economic and Political Weekly*, October 26, 1974.

Palit, S.B., 'Planning for better forests', *Yojana*, 20(12), January 15, 1976, p. 12-13.

Patel, V.J., 'Rational Approach towards Fuelwood Crisis in Rural India', *Economic and Political Weekly*, Vol. XX, No. 32, August 10, 1985.

Rao, M.S.A. (ed), 'Social Movements in India', Vol. 1 and 2, Manohar, New Delhi, 1979.

Report of the Scheduled Areas and Scheduled Tribes Commission, Vol. 1, 1960-61, Government of India, Manager of Publications, New Delhi.

Report of the Committee on Tribal Economy in Forest Areas, March, 1967, Government of India, New Delhi.

Report of the National Commission on Agriculture 1976, Part IX, Forestry, Government of India, New Delhi.

Report of the Development of Tribal Area, National Committee on the Development of Backward Areas, Planning Commission, Government of India, New Delhi, 1981.

Report of the Committee on Forests and Tribals in India, Ministry of Home Affairs, Government of India, 1982.

Report of the Committee for Afforestation of Barren, Denuded Lands and Hills of Maharashtra, Government of Maharashtra, 1982.

Report of 'Forests and Wood Section Conference', Indian Federation of Building and Wood Workers, D'Mello Bhawan, P. D'Mello Road, Carnac Bunder, Bombay-38, 1984.

Report of the Committee for the Review of Rights and concessions in the Forest Areas of India, Ministry of Agriculture, Government of India, New Delhi, 1984.

Role of Minor Forest Produce in economic and social development, Report of a Seminar held by Marathi Vidnyan Parishad, Udyam, Nagpur, 1986 (in Marathi).

Role of Women in Community Forestry, December, 1980, Proceedings, Ministry of Agriculture, Department of Agriculture and Cooperation, Forest Research Institute, Dehra Dun.

Roy Burman, B.K., 'Issues in Environmental Management Centering Forest, Role of Tribal Communities', *South Asian Anthropologists*, 1985, 6(1): 41-48.

Sagreiya, K.P., 'The Fodder and Grazing Position in Madhya Pradesh', *Indian Forester*, August, 1961.

——, 'Making The Most of Forests', Review of 'India's Forest Wealth by S.D. Thapar, *Economic and Political Weekly*, January 29, 1977.

Scenario of the 7 per cent, Vol. II, Cinemart Foundation, Development Area, New Delhi.

Shah, S.A., 'What ails Indian Forestry?' *The Times of India*, February 15, 1987, Bombay Publication.

Sharma,.B.D., 'Planning for Tribal Development', Prachi Prakashan, New Delhi, 1984.

Shiva Vandana, 'Forestry Crisis and Forestry Myths : A Critical Review of Tropical Forests : A Call for Action', World Rainforest Movement, 87, Cantonment Road, 10250 Penang, Malaysia (1987).

Singh, D.M., 'Forest Destruction Corporation of Maharashtra', Searchlighton Forest Department-series, *Nagpur Times*, (Daily), August, 1986, Nav Samaj Ltd., Ramdaspeth, Nagpur-10.

Singh, K.S. (ed), *Economies of the Tribes and their Transformation*, Concept Publishing Company, New Delhi, 1982.

——, (ed), *Tribal Movements in India*, Vol. I and II, Manohar, New Delhi, 1982, 1983.

——, *Birsa Munda and His Movements 1874-1901*, Oxford University Press, Calcutta, 1983.

'Social Forestry to the Benefit of big Farmers', Report from *Indian Express*, July 4, 1983, New Delhi.

Swamy, A.V., 'Forest Based Industries for Tribal Areas', *Khadi Gramodyog*, April 11, 1965.

'The Achievements of Full Employment in Wood Industries', Report I, II and III (1985). The Indian Federation of Building and Wood Workers, D'Mello Bhavan, P.D. Mello Road, Carnac Bunder, Bombay-38.

The State of India's Environment, 1982 and 1984-85, The Citizen's Report; Centre for Science and Environment, New Delhi, 1985.

'The Tropical Forestry : Action Plan', June, 1987, Prepared by FAO, World Resources Institute, The World Bank and the United Nations Development Programme.

'Tribals Hostile to Social Forestry', Report from *Indian Express*, July 11, 1983, New Delhi.

Tribal Research Bulletin, Tribal Research and Training Institute, Maharashtra State, Vol. III, November, 11 – March, 1981, Government of Maharashtra.

Tripathy Sangam, 'Forests and Forestry in India', Indian Federation of Building and Wood Workers, D'Mello Bhavan, P.D. Mello Road, Carnac Bunder, Bombay-38.

Vanjari, S., 'Socio-economic Conditions of Scheduled Tribes in Maharashtra', *Khadi Gramodyog*, April 11, 1965.

Various Plans for Tribal Development, Tribal Research and Training Institute, Pune, Maharashtra, 1984 (in Marathi).

Col. Verma, B.L., 'Vanishing Forests of Maharashtra', Water Resource Deptt., Community Centre, Development Area, New Delhi.

Vidyarthi, L.P., Sahay, B.N. (eds), *Applied Anthropology and Development in India*, National Publishing House, New Delhi.

Voelekar, J.A., 'Report on the Improvement of Indian Agriculture', London, Eyre and Spottiswoode 1893.

'West Bengal : When Forests Disappear, We will also Disappear', *Economic and Political Weekly*, November 27, 1982.

INDEX